W0115950

FLOURISHING IN THE AGE
OF CLIMATE CHANGE

FLOURISHING IN THE AGE OF CLIMATE CHANGE

William M. Throop

COMSTOCK PUBLISHING ASSOCIATES

AN IMPRINT OF CORNELL UNIVERSITY PRESS

Ithaca and London

Copyright © 2024 Cornell University Press

All rights reserved. Except for brief quotations in a review, this book, or parts thereof, must not be reproduced in any form without permission in writing from the publisher. For information, address Cornell University Press, Sage House, 512 East State Street, Ithaca, New York 14850. Visit our website at cornellpress.cornell.edu.

First published 2024 by Cornell University Press
Printed in the United States of America

Library of Congress Cataloging-in-Publication Data

Names: Throop, William, author.
Title: Flourishing in the age of climate change / William M. Throop.
Description: Ithaca : Comstock Publishing Associates, an imprint of Cornell
    University Press, 2024. | Includes bibliographical references and index.
Identifiers: LCCN 2023059434 (print) | LCCN 2023059435 (ebook) |
    ISBN 9781501777189 (paperback) | ISBN 9781501777202 (epub) |
    ISBN 9781501777196 (pdf)
Subjects: LCSH: Climatic changes—Social aspects. | Climatic changes—
    Psychological aspects. | Social change—Environmental aspects. | Climatic
    changes—Effect of human beings on. | Human beings—Effect of climate on. |
    Resilience (Personality trait) | Sustainability.
Classification: LCC HM856 .T476 2024 (print) | LCC HM856 (ebook) |
    DDC 304.2/8—dc23/eng/20240122
LC record available at https://lccn.loc.gov/2023059434
LC ebook record available at https://lccn.loc.gov/2023059435

For my grandson, Connor Torres Green, who I'm confident can flourish in these times

Contents

Preface

On beautiful fall afternoon a few years ago, I was meeting with a student who was failing my course on sustainability. His slumped shoulders and lack of eye contact communicated more than his responses to my questions. His motivation in the course was very low. As I asked about his goals and how he planned to pursue them, he turned toward me and vehemently exclaimed, "What is the point of doing all of this? We are all screwed!"

His outburst expressed what I was slowly recognizing has become almost common sense among a significant subset of young people. But not just young people. Many who care about sustainability and environmental issues share a similar feeling of pessimism. Declining biodiversity, failure to act aggressively on climate change, growing inequality, persistent racism, political polarization, and a growing threat of violence—all magnified by a crisis-oriented media—have led many people to think that the future is very gloomy if not apocalyptic.

Yet we cannot afford to allow pessimism, cynicism, and despair to undermine our commitment to sustainability or our pursuit of flourishing in the face of an onslaught of crises. For over a decade, my research focused on the social norms we must alter and the character traits we must strengthen if we are to live well in our times. Initially I aimed to identify what I call the "skillful habits" that would help people to change the world, but I gradually realized that I also needed to highlight the tools that would enable us to be highly resilient *no matter what happens in the future.* Ultimately, I reached the surprising conclusion that the skills we need to flourish in our troubling times are also critical for moving robustly toward sustainability. A win-win strategy emerged: If we cultivate character traits that maximize flourishing in the midst of major environmental and social decline, we will also strengthen traits that are crucial for shifting social norms toward sustainable solutions.

The task of flourishing no matter what happens should therefore appeal even to those who think "we are screwed." This book is written primarily for individuals, like my student, who are feeling hopeless, and for those whose professions involve motivating others to turn our collective trajectory toward sustainability. Both groups must confront the enervating power of pessimism about the future. In this book, I will share how each of us can cultivate the requisite skills and thereby find the motivation and the knowledge for how we can overcome our greatest challenges.

The most valuable lessons I have learned in writing this book is where I should strengthen my own skillful habits. Even where I thought I had practiced the requisite skills, I discovered significant blind spots. I have become better at collaborating with diverse peoples, less inclined toward problematic convictions, and better at finding beauty in my surroundings through writing about how such skillful habits contribute to flourishing. We know that we should develop the skills associated with our character throughout a lifetime, but we can all use reminders about where their further development could significantly improve our well-being. This book provides that guidance.

Acknowledgments

A book two decades in the making inevitably involves influential conversations and contributions too numerous to acknowledge appropriately. I am deeply grateful for the intellectual communities I have inhabited; they have stimulated and refined many of the arguments of the book. My colleagues at Green Mountain College willingly participated in many "test runs" of the ideas; in particular, Laird Christensen, Steve Fesmire, Heather Keith, Matt Mayberry, Tom Mauhs-Pugh, and Paul Stonehouse helped me significantly improve them. My students' reactions often led me to new insights. Rebecca Muffler and Kately Mann provided important research assistance.

Bruce Piasecki gave me an opportunity to present my ideas to business leaders and encouraged me to think much more broadly about my audience. At crucial moments, Mel Bringle helped me steer a new course that turned out to be far more effective. Steve Schwartz and Kimerer Lamothe made detailed comments on the entire manuscript and provided just the right mix of incisive critique and encouragement to guide the revision process. Two anonymous referees helped me to focus on the material that was most important, making the book much clearer. My mentor, Dick Prust, helped me polish the manuscript. My editor, Kitty Liu, and her able staff shepherded the manuscript to publication with patience and professionalism.

Early versions of a few parts of the book were published elsewhere, and I am grateful for being able to reuse this material. A different version of the main argument of the book can be found in "Flourishing in the Age of Climate Change: Finding the Heart of Sustainability," *Midwest Studies in Philosophy* 40 (2016): 296–314. The section on hope in chapter 2 is adapted from "Coping with Climate Despair: Cultivating the Skills of Hope and Tranquil Resolve," *Journal of Sustainability Education* 28, March (2023), coauthored with Paul Stonehouse. Parts of chapter 4 have been repurposed from "Frugality and Resilience: A Pragmatist Meditation," in *Pragmatist and American Philosophical Perspectives on Resilience*, edited by Kelly A. Parker and Heather E. Keith (New York: Rowman & Littlefield, 2019). Chapter 7 is an expanded version of "Learning Our Way toward Resilience," in *The Community Resilience Reader: Essential Resources for an Era of Upheaval*, edited by Daniel Lerch (Washington, DC: Island Press, 2017), 247–60.

My partner in all things, Meriel Brooks, has made writing this book possible in myriad ways. She has commented extensively on two versions of the manuscript and done an enormous amount of detail work where my patience would have failed me. She has been my north star and my joy.

FLOURISHING IN THE AGE OF CLIMATE CHANGE

INTRODUCTION

If we are to hope for the flourishing of all beings, every level of society, from the person to the family, from the neighborhood and municipality on up to the community of nations, must change.

—Stephanie Mills, 2021

Sometimes it seems we are all riding a roller coaster without brakes. While this may be exhilarating for a few, for most it is stressful and destabilizing. The speed of technological change, the rising rate of information flow, the increasing human impact on planetary processes, along with many other accelerating trends, create a context where flourishing is elusive. As large-scale threats multiply, fear and anger become commonplace, and dystopian stories abound. For many, the sense of decline is palpable. In a time of great change, we cannot assume that the skills reinforced by our culture will be best suited to our future. We try to change the world to make it conform to our preferred way of life, but the roller-coaster ride diminishes our sense of control. The less control we have over our circumstances, the more control we must have over ourselves—the more we need to adapt our character to the world as we find it.

The central argument of this book is deceptively simple: We face a distinctive set of challenges and opportunities during the first half of the twenty-first century, which I am calling the "age of climate change." Some character traits that are reflected in our cultural norms are ill-suited to our challenges. No matter what anyone else does, we will each increase our chances of flourishing in our times if we cultivate clusters of character traits that are in tension with strong cultural norms and hence underdeveloped. If many of us cultivate these traits, our likelihood of approaching sustainability increases tremendously, which will further enhance our flourishing. Thus, from a purely self-interested point of view, we should work on cultivating such traits, and if we care about the flourishing of others, we should redouble our efforts to strengthen these traits.

As a first step in unpacking this argument, I should explain what I mean by a few key terms. The age of climate change is characterized by an interlocking set of environmental, social, and economic challenges that reflect our rapid approach toward planetary boundaries. Environmental decline, reflected in biodiversity loss, ocean acidification, and climate disruption, is a major driver of the dynamics of this age, but income inequality, technological disruption, political polarization, rising threats of violence, and increased mass migration are salient features. Others have talked of the Anthropocene or the great acceleration, which are equally good candidates for naming the central dynamics of our times. I choose to highlight climate change in my characterization of the first fifty years of the twenty-first century, because it is more salient than other drivers of environmental decline. Its ecological and social impacts strongly influence other features of the period. I am not suggesting, however, that climate change is our most important issue or that if we solve it, we put an end to this period of ferment. Historians talk of the Gilded Age in late nineteenth-century America without implying that ostentatious wealth was its central issue.

To understand the age of climate change, we must understand the feedback loops between our rapid approach to planetary boundaries and our increasing social and economic volatility and conflict. Although we often experience this as an age of *cultural* decline, we need to see clearly the opportunities that characterize this period, as well as the challenges. This is the task of chapter 1, where I introduce resilience theory's adaptive cycle, which helps us to see how our challenges are interrelated and why we can expect more turbulence, whatever we do, as we "dwell on the edge of release."[1]

Flourishing through Emphasizing Different Skillful Habits

What would it mean to flourish in this age? Martin Seligman's account of flourishing is an excellent starting point for any discussion. He identifies five key elements of flourishing: positive emotions, deep engagement, strong relationships, a sense of meaning, and achievement.[2] All of these are involved in someone thriving or living well. Flourishing is not just being happy. Happiness is typically associated with subjective states such as the balance of positive and negative emotional states or the quantity of life satisfaction.[3] Flourishing, by contrast, is more closely connected with objective features that make life worth living. Aristotle's theory, according to which one achieves eudaimonia by developing human excellences, is an early example.[4] The capabilities approach to well-being is a more recent version.[5]

People are diverse enough that it seems unlikely that we can specify a unique set of defining characteristics of a flourishing life. Flourishing, then, is best understood as a cluster concept. We can specify a cluster of features of a life that are associated with flourishing across many contexts, and we can be confident that if one has enough of the features in this cluster, then one is flourishing.[6] Seligman's list includes subjective elements like positive emotions, but it also contains objective features like strong relationships. I would augment Seligman's list with three other elements. To flourish, we need to have our basic needs met, such as food, shelter, and bodily security.[7] We also must have the capability to be self-directed. Lastly, we must have some understanding of the forces shaping our lives and some reasonable confidence in our capacity to navigate those forces. This last element is especially germane to my claim that we need to strengthen key skillful habits that help us to flourish in our times.

We could continue to add elements of flourishing to the list, but most would specify further features identified abstractly above. Individuals could flourish without any one feature listed, but unless they are quite unusual, their flourishing decreases if they lack multiple features on the list. Prudent people will try to understand both the local and global contexts that are likely to shape their lives. They would then try to develop their capacities to flourish in the likely range of contexts they may face. If my central argument is right, for most of us such prudence requires significant change in our dominant character traits.

But why focus on character? Isn't the best way to foster flourishing to shift our policies and adopt better technologies? About fifty years ago, the deep ecology movement was given its classical formulation in Arne Naess's famous essay "The Shallow and the Deep, Long Range Ecology Movement. A Summary."[8] Naess juxtaposed a shallow, technological and anthropocentric approach addressing environmental problems with a deeper approach that required fundamental ethical change. At the time, Julian Simon and others argued that free markets and technological changes were sufficient to deal with resource depletion and waste. Many environmentalists focused on policy changes and economic incentives, but deep ecologists went further, maintaining that we need to change our sense of who we are and our relationships to the biosphere.

Naess was primarily focused on the question of what changes were necessary to create a positive relationship between humans and nature and to reverse environmental decline, not on the question of how we can individually flourish. But he thought his solution to environmental decline would enhance individual flourishing. My priorities are the reverse. I emphasize the question about individual flourishing but argue that my answer will be necessary to address our other significant challenges. Deep ecology is rarely talked about in today's environmental circles, but the issue Naess raised about how deep our changes need to be remains very much alive. I have reservations about the specific changes he recommended, but I agree that we must change who we are, how we see ourselves, and how we relate to others. Throughout the book, I will show why shallow changes are unlikely to be sufficient for either our individual flourishing or our global approach toward sustainability.

Character and Flourishing

The connection between character and flourishing has a long history. In the West, Aristotle, Stoics, and Christians saw cultivation of character as a crucial part of a flourishing life. In the East, Confucianism and Taoism arrived at similar conclusions while focusing on rather different traits. Today, positive psychology shapes global discussions of ways in which character is linked with flourishing. Peterson and Seligman provide an overview of twenty-four character strengths that are integral to living a good life.[9] I agree that many elements of character are linked to flourishing, but I suggest that the systems dynamics of our context and the cultural norms we inherit determine which elements are sorely in need of development. I will focus on four clusters of traits that are underemphasized in the US and other Western countries and that are particularly important for our flourishing now. Two of these, humility and frugality, are normally considered character traits, but the other two, systems thinking and collaboration, are more often thought of as sets of skills.

A character trait is a relatively stable pattern of thought, feeling, and action that is seen as part of one's identity.[10] Many common character traits, like honesty, courage, and generosity, descend from theories of virtue and vice inherited by our cultures. We often lack terms for elements of character that are distinct from that inheritance. We have no common names for many stable clusters of skillful habits that are part of our identities and that otherwise perform the function of character traits. Virtues as classically understood are not just dispositions to think and act in various ways; they involve many skills, including subtle judgments about what a virtue requires in a situation and how that is affected by other relevant virtues. I will typically use "skillful habit" for the elements of character I am recommending we cultivate, thereby emphasizing that we are really looking to develop skills that become a part of who we are—cognitive skills, emotional skills, and behavioral competence.[11] To be clear, I am examining very general skill sets that are deemphasized by a culture, not specific skills that might be in demand in a twenty-first-century marketplace, given economic trends.

Speaking of some character traits as skillful habits has several advantages. Skills are typically individuated much more finely than character traits, so talk of skillful habits multiplies the elements of character, which expands our sense of who we are. Furthermore, skills are usually learned by breaking them down into subskills, which we can practice individually. To become a better writer, I might focus on specific skills, for example paragraph construction or sensitivity to audience. Similarly, becoming a more collaborative person might involve honing subskills that are underdeveloped. Not all skillful habits, however, are broad and significant enough to be aspects of character—think of the skillful habits involved in riding a bicycle. Moreover, not all character traits involve skills; some are just dispositions to behave in certain ways, like being crass. Both dispositions and skills are important for the kinds of traits I am discussing. If I am a highly collaborative person, for example, then I am not just inclined to be collaborative; I am also good at it.

Each of chapters 2 through 5 describes a cluster of skillful habits that we need to strengthen if we are to flourish *and* if we are to approach sustainability. Chapter 2 explains why strengthening the three collaborative skills of empathy, trust, and hope is particularly important amid the conflicts of this period. We often think of these three as matters of affect rather than skill, yet the skills they involve are crucial capacities for working together to create livable communities now. Our culture's strong emphasis on competitive skills makes it hard to reinforce collaborative skills. Nevertheless, competitive skills remain important. I argue we should achieve a kind of "binocular vision" that integrates the use of collaborative and competitive skills while emphasizing the former.

When competitive skills dominate a culture, social norms reinforce conviction rather than humility. But humility is crucial for learning about the dynamics of our age and working together to moderate their effect. Chapter 3 describes how powerful "conviction conveyor belts" in our culture turn ordinary beliefs into hardened convictions. By cultivating humility skills, we can resist this tendency. In times of rapid change and novel challenge, humility skills are necessary for the learning mindset we need to flourish.

We live in a time of abundance unimaginable to our ancestors, and yet we do not appear to be much happier than they were. A history of technological achievements leads us toward a faith that technology will solve most of our problems, which reduces our motivation to pursue the cultural changes more likely to make us happier, including development of frugality skills. Chapter 4 builds the case for a broad set of frugality skills that are necessary in an age of increasing resource constraints and system dysfunction. Such skills enable us to reduce our consumption without loss of well-being, which is crucial for approaching sustainability.

Chapter 5 argues for remediating deficits in our cognitive skills that make it harder for us to fully grasp the dynamics of our times. I describe a cluster

Current cultural norms support this skills distribution

These skillful habits have helped us get to where we are today.

Dominant	Secondary
Competitive Skills	Collaborative Skills
Conviction Skills	Humility Skills
Abundance Skills	Frugality Skills
Individualist Skills	Systems Skills

Paradigm shift

Secondary	Dominant
Competitive Skills	**Collaborative Skills**
Conviction Skills	**Humility Skills**
Abundance Skills	**Frugality Skills**
Individualist Skills	**Systems Skills**

These skillful habits will enable us to reduce risk and grasp opportunities.

Skills for flourishing in our times

FIGURE 1. The shifts in skillful habits that will increase the odds of flourishing in the age of climate change.

of systems-focused skillful habits that must be strengthened if we are to confidently navigate these dynamics. Our contrasting individualist cognitive habits, which served us well enough during a period of massive growth, now make it harder to effectively meet our large-scale challenges. In each chapter, I am recommending a change in emphasis and expertise regarding underdeveloped skillful habits, not the wholesale repudiation of contrasting skills. Figure 1 illustrates these changes.

Cultural Change and Sustainability

I will call my four clusters of skillful habits "resilience" skills or traits because, taken together, they strengthen both personal resilience and the resilience of our communities. Here resilience means roughly the ability to successfully navigate a wide range of challenges.[12] The salience of trauma and adversity has made research on psychological resilience a major growth industry. Numerous skills have been associated with enhanced individual resilience,[13] only some of which are encompassed by the four clusters I emphasize. Personal and community resilience are often treated separately, but I think it important to focus on the feedback loops between them. As we will see, finding groups that are characterized by these resilience traits and that positively reinforce their further development plays an important role in flourishing.

In chapters 6 and 7, I turn to the challenge of how we can scale up cultivation of resilience skillful habits so they begin to be reflected in cultural norms that are necessary for us to approach sustainability. I provide a full account of sustainability as a social goal at the end of chapter 5 when we have reviewed the systems language that the account requires. In the meantime, we can say with Agyeman, Bullard, and Evans that sustainability "ensures a better quality of life for all, now and into the future, in a just and equitable manner, whilst living within the limits of supporting ecosystems."[14] So understood, sustainability clearly involves social and economic elements, not just environmental health. In particular, it requires justice both for humans and nonhuman organisms.

Major shifts in our cultural norms are necessary for us to approach sustainability, but they are certainly not sufficient. We also need technological, policy, and economic shifts. But without a shift in norms, these other changes are very unlikely to be effective and stable over time. This kind of major adaptation is very hard to achieve in the best of conditions. Many are resisting the requisite changes and cannot see how adapting our character is compatible with leading a fulfilling life. Sustainability is a long-shot goal, but even if we fail at this goal we can still flourish amid the resulting turbulence.

Caveats

Most readers will have a variety of questions and concerns about my project. For example, it might seem highly implausible that we can flourish during our transition. Perhaps the most we can do is to cope with the fallout and live minimally decent lives. Or perhaps flourishing will be limited to a few elites, whereas ordinary people living on the edge of survival cannot expect to flourish. These are legitimate concerns. Clearly people who cannot meet their basic needs cannot flourish. But my claim is that most people across the socioeconomic spectrum can flourish even in these times. Humans have flourished in a very wide variety of contexts, including those where things are going badly wrong. Without wealth and advantage, we can learn how to build strong, nurturing relationships. We can find ways to be deeply engaged in activities that enhance some corner of the world and give us meaning and accomplishment. We can find beauty and other sources of pleasure that stimulate positive emotions. And we can learn how to understand and navigate effectively local systems that directly affect our lives.

To be sure, if we do not rapidly move toward just sustainability, many more people will not flourish because their basic needs will not be met. Alas, many more of those who can meet basic needs may not learn how to adapt, because their cultures may thwart such learning. They too will suffer. Much more needs to be said regarding these concerns, and that is one aim of the book. I will address different versions of these concerns directly in chapter 6, which explores the barriers to developing the resilience traits individually and as a society. If these barriers are steep, then the chances of scaling up the requisite resilience traits are low. I argue that many of the barriers to individual cultivation are small, and I describe a range of stepladders that will help us surmount the barriers. This chapter also serves to clarify and augment the picture of how, on a personal level, each of us can buck social norms and strengthen resilience skillful habits.

The barriers to fostering widespread development of resilience skills are much higher and the stepladders more rickety, yet much is happening that suggests success is within our grasp. These skills have strong historical roots in our cultures that are motivating to conservatives and liberals alike. From indigenous peoples and communal enclaves to the social uprising that Paul Hawken calls "the movement that has no name," we see groups that positively reinforce the relevant skills. What we need now is major institutions that can unify, expand, and make more visible the practices in these groups that cultivate the skillful habits we need.

In chapter 7, I argue that education, broadly understood, could provide the institutional support necessary for scaling up resilience traits, but to do so, it would need to change considerably. While some initiatives in K–12 and postsecondary education are moving us in the right direction, formal education alone affects a relatively small portion of society in the short run, the younger generations. We also need nonformal education, especially in businesses and religious organizations, to reach a broader segment of the population with offerings and cultural norms that reinforce resilience skills. Here again we see significant momentum among these institutions, but also polarizing forces that limit their impact. Culture change can occur with surprising speed if the conditions are ripe. We are justified in allocating a great deal of hopeful energy to this enterprise.

In the epilogue, I sketch two scenarios: one in which resilience skillful habits are manifested only in small enclaves within a world where deep conflict dominates, and the other in which they are widely embraced and conflict is managed more effectively. Some intermediate scenario is more likely, but my point in the exercise is to show vividly that in either scenario we are more likely to flourish if we have a strong set of resilience skills. While far from a panacea, these skillful habits enable us to better navigate the turbulence that we can expect on the edge of release.

Throughout the book, I liberally refer to "we" as if it were clear whom this includes. For two reasons, I will focus my discussion primarily on the range of people living in the United States, though much of what I say will apply with suitable modifications to people living in other postindustrial cultures. First, I know more about the United States than other countries. It is hard enough to generalize productively about the cultures of any large country given the fragmentation of subcultures in the modern world, but it would be even harder to generalize about larger groups of cultures across the globe. The data available about the United States have spawned a robust literature about US cultural character and its changes, which has guided my thinking. Second, the United States has played an outsize global role in the last century both economically and socially, which resulted in its exporting many culture norms. Moreover, many of the forces that have influenced general social norms in the United States such as capitalism, the Protestant revolution, and the evolution of modern science, have also influenced much of the globe in similar ways. Consequently, arguments I am making about US culture should resonate in many other cultures. The United States has been a laboratory for refining skillful habits associated with individualism, competition, conviction, and technological abundance. Our social norms strongly reinforce such skills and de-emphasize contrasting skill sets. The country can now serve as a different kind of laboratory in which we experiment with the large-scale cultural change necessary for widespread flourishing.

Some will still have doubts about how far we can generalize across the variety of subcultures in the United States. I am mindful that many subcultures do not fit my generalizations about our dominant cultural norms. Despite the tremendous diversity of people, values, and personal situations we find in the United States, we can see common themes that apply to large segments of the population and that affect how we as a whole national body respond to our challenges. Even if one inhabits a subculture characterized by norms that reinforce sustainable living, one is affected deeply by the norms that dominate the larger culture. Some are already on the path I am recommending, but we can all strengthen the skillful habits best suited to flourishing in our times and help others by contributing to changes in our cultural norms.

To see why resilience skillful habits are likely to enhance our own flourishing no matter what ultimately happens in the age of climate change, we need to understand the dynamics of our times. That is the task of the next chapter. We need to see the interconnections between various challenges *and also* the opportunities that these provide. Here we begin to see how the pursuit of individual flourishing, even if we think that we are screwed, can align with making progress toward sustainability.

THE PROBLEM OF FLOURISHING IN OUR TIMES

> The world is going to end in 12 years if we don't address climate change.
>
> —Alexandria Ocasio-Cortez, 2019

Are we on a precipice gazing toward catastrophe? In a recent book, David Wallace-Wells argued that we are on track to make large parts of our planet uninhabitable.[1] James Kunstler projected that "we are entering an era of titanic international military strife over resources."[2] Byron Williston argued that "it is imperative that the catastrophic framing mode moves from the environmental fringe to the cultural mainstream," and he envisions the possibility that justice may not even be possible for future humans given severe resource scarcity, or it may involve a survival lottery.[3] Greta Thunberg maintained that "around 2030 we will be in a position to set off an irreversible chain reaction beyond human control that will lead to the end of our civilization as we know it."[4] Dire projections have been a staple of environmentalism for more than fifty years. With our failure to adequately address climate change and other environmental problems, such claims have become increasingly common and dramatic.

Flourishing would be impossible for most people if such apocalyptic projections were true, but most of them are exaggerated. At best, they highlight worst-case scenarios that could occur *if we did not make any changes in our current practices.* But relevant changes are happening—though too slowly. Time and again, we do act to mitigate crises, though often too late to avoid serious consequences. In the 1970s, dire projections about human population growth leading to mass starvation were avoided because of the green revolution in agriculture, though this caused other problems.

To be fair to Ocasio-Cortez, Thunberg, and others who make apocalyptic claims, their aim is admirable. Such claims are used to jar us out of our sense of normalcy and strike enough fear in us that we make radical change now. Thus, they do not think that catastrophe is inevitable if we act now. Alas, the tactic of using fear to stoke action often backfires. As Michael Shellenberger argues in *Apocalypse Never*, doomsday scenarios can make our challenges seem so over-whelming that we give up hope and turn to living in the moment. They can also further polarize people into insular camps that focus on fighting each other, and it creates cynicism about scientific projections, especially when the apocalypse fails to occur.[5] Even when we realize the aims of such claims, they tend to erode our confidence that we can surmount our challenges, and thereby increase our anxiety.

I agree that we do need to change much of our behavior, and that up to now, we have failed to address our daunting challenges adequately.[6] So am I just disagreeing about tactics? Not entirely. Because of the work of a great many people, we *are* making significant progress on climate change and other issues. This progress will probably accelerate, despite the fierce backlash it has gener-ated. As a result of all this activity, we will probably avoid civilization collapse, but we will face a highly turbulent future for the next several generations as we navigate accelerating systems change. We face a future of half-measures and heightened conflict, of solutions too late and too little, but a future where few outcomes are inevitable, where we still have options. What we do matters a great deal.

We need a different approach to motivating the massive behavior change in the direction that will enable broad-based human and nonhuman flourishing. I doubt that either fear or a shared positive vision will be broadly effective. Even though in some contexts fear does work as a motivator, it contains significant risks. Fear can shut too many people down, and it engenders skepticism and distrust.[7] Achieving a widely enough shared positive vision seems highly unlikely in our polarized society. The most promising approach to motivating large-scale behavior change is to identify places where pursuit of individual self-interest aligns with the common good and to highlight the promise of the win-win for promoting flourishing. Cultivating the skillful habits best suited to our chal-lenges and opportunities fills that bill.

A Stark Summary of Our Challenges

Most of us are acutely aware of the environmental, social, and economic chal-lenges I summarize below. But often we think of them in isolation and fail to see the feedback loops that prevent them from being fruitfully addressed in piece-meal fashion. We also tend to focus on what we are likely to lose rather than what we can gain from addressing these challenges. We underappreciate the opportunities that lie within them. This chapter aims to rectify these deficits. We need a holistic vision of our challenges and opportunities to motivate cultivating the skillful habits necessary for flourishing now.

Let's begin with our environmental challenges. A defining challenge of these times is that we are approaching key "planetary boundaries," limits to anthropogenic changes in earth systems that support human societies. The research on planetary boundaries synthesizes scientific work on earth system processes that enabled human civilizations to arise, especially those we risk destabilizing.[8] The boundaries delineate what we know to be a "safe operating space" for the earth system. Researchers argue that we appear to be crossing boundaries for biodiversity loss and nitrogen/phosphorus pollution, while we are approaching them for climate and land-use change. The boundaries do not necessarily consist of tipping points beyond which there is no return, though that may happen. They indicate points beyond which it will be harder for civilizations to flourish. We can expect increased turbulence in earth system processes as we approach these boundaries.

Climate disruption is one of the large drivers of other changes in the earth system. It affects biodiversity, ocean acidification, desertification, and agricultural productivity. It contributes causally to an increase in "natural disasters" like flooding, fires, and droughts. Human population growth and consumption levels are the other main drivers of shifts in planetary processes. Although the rate of global population growth is slowing, the sheer number of people continues to increase dramatically—approximately eighty-two million per year. Population growth contributes to transformation of wildlife habitat for food and other resources. The resulting deforestation further exacerbates climate change, and widespread pollution contributes to further biodiversity loss. Our consumption patterns magnify the effects of population growth—creating tremendous waste, increasing destruction of habitat, and fueling grossly unjust distribution of goods and harms.[9] The interactions among these challenges make it very hard to effectively address one without addressing others. Our standard problem-solving approaches favor focusing on a single issue, but holistic approaches are necessary to alter feedbacks between different issues.

Alas, these large-scale challenges seem remote from daily life. Because of their global nature, planetary boundaries are not readily apparent, rendering these challenges invisible. Their impacts depend a great deal on our specific situations. The wealthy have been able to avoid some negative impacts. The poor and marginalized bear the brunt of our failures to meet such challenges. This pattern will continue, but increasingly everyone will bear significant costs.

We must make the personal impacts of our large-scale challenges more salient. The following is a representative list of the ways that approaching planetary boundaries affect human flourishing, directly or indirectly:

- The rising severity and number of natural disasters create greater insecurity, supply-chain disruption, and increasing governmental costs and taxation.
- Shortages of food and water lead to more social conflict and increased cost for necessities.
- Human migration is increasing, and with it resistance to immigration and militarization of borders.
- Biodiversity losses destabilize ecosystems, leading to loss of pollinators, increases in pest species, decline of coral reef ecosystems, and increased expense for ecological restoration.
- Those who understand our situation are beginning to experience a growing sense of sadness about the relatively stable nature we are losing—what some have called "solastalgia."[10]

This summary of environmental challenges highlights some of their social impacts, but the age of climate change also includes distinctive social challenges. These include declining trust in institutions and each other, increased sectarianism, aging populations, rapid social change, highly salient injustices, and rising authoritarianism. The ecological challenges arising from approaching planetary boundaries are leading to increased social conflict around how to address boundary issues. Historic injustices influence how we distribute the costs and benefits of avoiding planetary boundaries and thus exacerbate conflict. Tensions between developing and developed nations and between the beneficiaries of global capitalism and those harmed by it threaten to create feedback loops that reinforce our ecological challenges.

Our 24/7 media are major drivers for our social challenges. We are ensconced in a media environment that has an insatiable demand for controversy. The economics of most media incentivize shrill and often misleading headlines that attract more article views. Social media platforms tend to prioritize news feeds that reinforce our preferences, and increasingly we are divided into insular media environments that reinforce polarization. The media-fueled enmity reflected in the polarization around climate change and the COVID-19 pandemic illustrates how our social dynamics increase the difficulty of effectively addressing other challenges. Such dynamics decrease shared trust in institutions like government, science, and the press and erode the social touchstones that build unity across disagreement.

In the US and most of the globe's largest economies, an aging population creates economic problems and makes it harder to address other social problems. The number of people in the United States over the age of sixty-five will double between 2012 and 2050, while total population is projected to grow only 27 percent.[11] The challenges resulting from an aging population include increasing government health-care and retirement costs, which deprive other initiatives of funding and enlarge government debt. This decreases the capacity of governments to address social needs, which in turn reinforces distrust in institutions.

These large-scale social challenges have a host of impacts on individual flourishing:

- We experience increasing anger at those who seem to thwart our progress and despair about achieving negotiated solutions to problems.
- Fear of intractable conflict and physical violence is increasing.
- Mental health problems have been on the rise, and that is likely to continue.
- Most people except for social elites will reasonably see that system structures are rigged against them, and even elites will feel like victims of system dynamics beyond their control.
- We have a heightened sensitivity to the speed of change and a diminished sense that we can control our destiny.
- Compared to the past, our social support structures tend to consist of smaller numbers of close friends and family and larger groups of acquaintances with whom we have only thin connections. Loneliness becomes a persistent problem.
- Institutions like families, schools, and governments often cannot manage the demands that are placed on them, which decreases their effectiveness and our trust in them.

Our economic challenges reinforce our social challenges, and vice versa. Our dominant economic story remains one of material progress, but many are losing confidence in rising material standards of living. Our expectations have been conditioned by seventy-five years of extraordinary economic growth that appears unsustainable given our ecological constraints and our experience of the great recession of 2008. The brutal competition of a globalized capitalism and the speed of technological change are twin drivers of economic insecurity. Globalization enabled many industries to shift factories to countries with cheaper labor and to contract internationally for services, which severely reduced well-paying jobs for those without baccalaureate degrees and weakened the bargaining power of labor.

The rate of technological innovation continues to accelerate, which provides many benefits but also spreads economic disruption across a wide range of industries. Automation and artificial intelligence threaten many jobs that once seemed like secure careers. Long-term employment in a company is increasingly uncommon, and young people are told that they will likely have more than seven distinct careers. The gig economy in which people work for numerous employers as independent contractors has started to dominate some economic sectors. Uber drivers, adjunct faculty, and writers for hire can manage their own time and workload but must do so without security, workplace health insurance, or retirement benefits. The primary beneficiaries of the above drivers are shareholders and managerial elites. As fewer people garner these benefits, the vast majority feel at the mercy of external forces that cause job loss, unsustainable debt, devaluation of once marketable skills, and social displacement. Increasing inequality, global interdependence, and the difficulty in a low-growth economy of achieving high employment levels with reasonable wages underlie the fragility of our economic situation. Our polarization prevents us from jointly crafting economic changes that enhance job security and support workers whose livelihoods are threatened by technological change. As a result, our primary political approach to addressing economic insecurity is to stimulate high economic growth and low unemployment, which fuels the consumption that erodes our planetary buffers.

The dynamics of our economy lead to many individuals experiencing the following personal challenges:

- We work harder and feel less rewarded. A chronic sense of unfairness grows.
- The economy demands more education, which currently leads many students into unmanageable debt.
- Too many of us struggle to find meaningful work that pays a livable wage and enables a dignified retirement.
- The speed of technological change places a premium on novelty, which reduces the reward for steady hard work.
- Increased economic insecurity makes us suspicious of those who might encroach on our jobs or work for lower wages.
- Our health suffers from the stress of insecurity and low job satisfaction.

But doesn't every generation have its problems? Aren't all these challenges best understood as problems that we need to solve? Almost fifty years ago in a now famous article, Rittel and Webber drew a distinction between "tame" problems and "wicked" problems.[12] Tame problems may be quite complicated, but they have definitive problem formulations, and eventually experts can discover clear solutions for them. Engineering problems like designing an electric car that can travel three hundred miles on a charge fit this category. By contrast, wicked problems have contested problem descriptions; often stakeholders define the problems in contrasting ways, which fit with their preferred solutions. There are no stopping rules that indicate when a solution has been reached; no amount of new data or refined algorithms will determine how to solve the problems. Indeed, wicked problems tend to involve complex social and ethical issues about which people with conflicting values will reasonably disagree. Such problems are often interconnected with other problems and are highly context dependent. They have better or worse solutions at a time, but not correct solutions. Climate change, structural racism, and political polarization are wicked problems.

We have been so successful solving tame problems that it is natural to assume that social problems can be treated analogously once we have enough data about how people behave. Behavioral psychology and classical economics seemed to promise to tame social problems, but alas that promise cannot be realized because of the variability among humans interacting with the complex socio-ecological systems they inhabit. Most of our important challenges are wicked problems that can only be managed in ways that are temporary, contested, and context specific; they will not be "solved" in the traditional sense. We are stuck with them or the negative effects of dealing with them.

By highlighting our challenges above, I have inevitably created a bleak picture of the times in which we must try to flourish. But that picture is also very one-sided. Tremendous opportunity accompanies the disruptions we face. We also have the benefit of the many forms of progress that have characterized the last hundred years. We live in a time of great change, which opens up avenues for creativity that in other ages might have been less available. The importance of addressing these challenges opens many opportunities for living a life full of purpose. The cumulative efforts of many will determine how we collectively move through the turbulence of our times. We cannot know whether the net result of our activities in this century will lead toward a sustainable and just society or toward some less desirable alternative, but we do know that what we do matters a great deal. Before turning to a more detailed treatment of the opportunities we might grasp, I describe a model that provides a picture of how these challenges and opportunities fit together.

Navigating the Adaptive Cycle

If we try to address the above challenges in a piecemeal fashion, we become overwhelmed by their number and our difficulty working through wicked problems. We need a unifying picture that helps us to understand the complex dynamics of our times and leverage large-scale change that addresses many problems simultaneously. I have found that resilience theory's model of the adaptive cycle provides such a picture *and* reveals the kind of opportunities available in our times. It also illustrates why our cultural norms might not be well adapted to our challenges and why it is important to strengthen different skillful habits. This picture provides a promising explanation of why after a long period of rapid growth, our accumulated resources and institutions have become brittle and ineffective— why we need massive change, and why it is so difficult to achieve.

As illustrated in figure 2, the adaptive cycle consists of four phases that complex social and ecological systems typically go through—growth, conservation, release, and reorganization. Once we see how these phases work in a simplified ecological example, we can move to a more complex social example that better reveals the structure of our current situation.

The adaptive cycle arose out of ecology, and its four phases are most easily explained by means of a simplified biological example.[13] In the growth phase (r), resources are readily available, and organisms compete intensely for them. The structure of the system becomes more complex, but it is still highly flexible, adapting quickly to changes in its external environment. Consider a fire-prone ecosystem like Yellowstone National Park. After a forest fire, early succession species grow quickly where their seed stock has survived in the soils. The available energy fuels intense competition for sunlight, nutrients, and water. After a while,

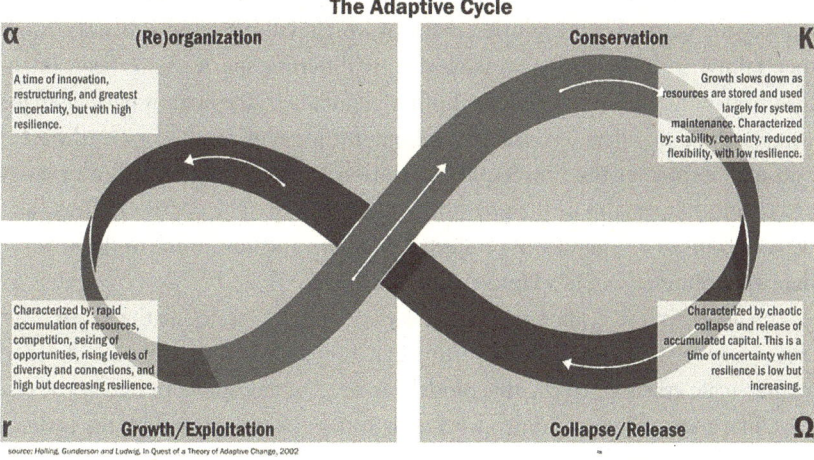

The Adaptive Cycle

α   **(Re)organization**   **Conservation**   K

A time of innovation, restructuring, and greatest uncertainty, but with high resilience.

Growth slows down as resources are stored and used largely for system maintenance. Characterized by: stability, certainty, reduced flexibility, with low resilience.

Characterized by: rapid accumulation of resources, competition, seizing of opportunities, rising levels of diversity and connections, and high but decreasing resilience.

Characterized by chaotic collapse and release of accumulated capital. This is a time of uncertainty when resilience is low but increasing.

r   **Growth/Exploitation**   **Collapse/Release**   Ω

source: Holling, Gunderson and Ludwig, In Quest of a Theory of Adaptive Change, 2002

FIGURE 2. The adaptive cycle. Adapted from *Panarchy*, edited by Lance H. Gunderson and C. S. Holling. Copyright© 2002 Island Press. Reproduced by permission of Island Press, Washington, DC.

young trees begin to lock up more energy in the system; the forest begins to return. Numerous dead trees and branches drop to the ground, providing cover for animals and also providing the tinder that can fuel another fire.

When the forest matures, its growth slows, and large amounts of energy are stored in the biomass of the forest. Now the ecosystem has entered the conservation phase (K). Much of the energy available to the system (e.g., sunlight and water) is used to maintain its structure, which tends to become increasingly rigid. The system becomes highly interconnected but also less flexible. Despite occasional disturbances like a wind event or an insect invasion, the forest grows slowly and retains its identity, which is defined in terms of stable processes that maintain its structure. It becomes an old-growth forest.

In the absence of frequent fires, the forest litter builds up and makes stand-changing fires more likely. Eventually, a large disturbance may push the forest over a threshold in which it loses key structures and functioning. A bolt of lightning hitting a dead tree can then lead to stand-changing fire—a crisis from which the forest cannot rapidly bounce back. The system has now entered the release phase (Ω), where energy locked in the wood is released back into the system as heat, ash, and gases. In this phase, the feedback loops that maintained the old structure are no longer operative. Organisms that survive the fire now interact with a novel environment. Release can be fairly shallow, where the system retains some of its prior identity, or it can be very deep, where the prior structure and processes collapse. In either case, rapid chaotic change becomes the norm for a time.

In the reorganization phase (α), clusters of organisms establish themselves, forming novel small communities, many of which will fail. Experimentation and uncertainty are high and success elusive. If different species have been introduced since the last major fire, or if other elements in the system have changed (e.g., much of its soil has eroded), the reorganization may result in a very different kind of system after the fire. Rapid innovation and competition will ultimately lead to another version of the growth part of the cycle. The depth of release and the variations in reorganization trajectories determine how much of the system's identity is retained as the adaptive cycle continues.

The adaptive cycle plays out on larger and smaller scales at the same time. Large-scale systems tend to move through the cycle much more slowly than smaller-scale systems. To use this model, we need to envision a nested set of systems, like a set of Russian dolls, with each system being affected by the systems that it includes and by larger systems that include it. For example, a changing climate will affect which tree species will do well after a forest fire in a region. A developing theoretical framework explains the dynamics of the adaptive cycle as it applies in different settings. When the pattern of growth, conservation,

FIGURE 3. The adaptive cycle applied to social systems. From Brian Fath, Carly Dean, and Harald Katzmair, "Navigating the Adaptive Cycle: An Approach to Managing the Resilience of Social Systems," *Ecology and Society* 20 (2015): 24.

release, and reorganization is applied to social systems—human communities or organizations—it must include both the intentions and the skills of community members. Figure 3 shows a more complex version of the adaptive cycle as applied to a socio-ecological system.[14]

In this diagram the phases are labeled somewhat differently, but we can see how the ecological model is adapted to social systems like organizations and communities. The oscillating line moving through the growth and conservation (status quo ante) phases illustrates that small systems within an organization are going through their own adaptive cycles while the organization as a whole grows; progress is never linear. For example, a new sustainability studies department at a college may grow rapidly, assisting in the overall growth of the college, while the chemistry department shrinks.

If an organization grows enough, it begins to standardize practices, hire more middle-management staff, and build structures for solving routine problems. It gradually enters the conservation phase. It builds up reserves that help maintain its structures through minor crises, but at the same time the growing bureaucracy can become rigid and less capable of adapting to external pressures. Decision-making processes can become cumbersome, and competing groups of stakeholders can make it hard to solve problems. Often, though, the conservation phase is long and productive, stimulating incremental changes that keep the organization functioning well and that create pride and high morale. For example, the iconic film and camera company Kodak moved through a rapid growth phase in which it introduced the first inexpensive camera, the Brownie, which made cameras into household items. It created a huge market for its film and film-processing chemicals. Kodak had over a fifty-year run in the conservation phase, innovating and creating products people admired. The company even invented an early digital camera, but digital competitors eventually innovated more effectively, and Kodak did not fully anticipate the implosion of the analog film business. Once it lost the film revenue, it was on the way toward release. Kodak declared bankruptcy in 2012.

A key threshold for an organization like Kodak is having enough revenue to innovate in ways that generate future growth. Once the larger-scale system moved toward digital photography, Kodak was caught in a downward spiral. As figure 3 illustrates, a strong organization in the conservation phase can handle many crises without crossing a major threshold. But if it is weakened enough that a crisis pushes it across a threshold, it enters release. Now it loses structures that would enable it to bounce back quickly. This chaotic phase often involves loss of faith in leadership and rapid changes in hierarchies of authority. New leaders must arise to fill the vacuum, and new relationships must be built within the organization as old hierarchies become irrelevant. Release also involves loss of standardized procedures and a much greater reliance on improvisation.

In apocalyptic environmental literature, release is often associated with total collapse or transformation into a dystopian nightmare.[15] While this is a possible result of crossing a major threshold, it is not the most common consequence, Often, release is less severe, and parts of the prior culture are preserved but in a different form. Even where a culture dissolves, its people migrate elsewhere and join other cultures. My disagreement with the apocalyptic framing of our situation is about the probable consequences of a major release, not about the likelihood of some release occurring.

Indeed, release is not always bad, though it typically involves a great deal of uncertainty and stress as standard rules and roles break down. In a dysfunctional organization, where incremental positive change has been systematically blocked, a crisis that pushes it into release may be a necessary prelude to reorganizing with a healthier organizational culture. A global pandemic, an economic depression, and a ruthless authoritarian regime can be very resilient and thus hard to change. In such systems, we may deliberately push a system over a threshold into release in order to create space for reorganization. In such a case, resilience theorists describe the release and reorganization process as a system transformation. Often sustainability is viewed as prolonging the conservation phase indefinitely by keeping a valuable system from going over a key threshold, but some argue that sustainability requires system transformation that cannot be achieved without going through release and reorganization. In general, the more fundamental the changes one seeks, the more likely the system needs to go through at least a mild release and reorganization.

The reorganization or "innovation" phase (as it is labeled in figure 3) in an organization involves looking for a different way of operating that works effectively within the larger context and has the potential to lead toward a new stable pattern of growth. Here entrepreneurial activity is very important. The freedom to experiment is crucial, but if it is carried on too long or if insufficient investment is made in promising experiments, an organization can become stuck in the reorganization phase and never return to growth. Kodak appears to be in this situation. Its bankruptcy was not the end of the company. It retained its movie film and chemicals business, which continue to be somewhat profitable. It is a much smaller company today than it was twenty years ago, however, and it has not yet found a new growth engine.

This very brief overview of the adaptive-cycle model gives us a glimpse of how complex systems are likely to behave over time. Although it is difficult to know for sure how to apply the model to a specific context, it is very helpful to estimate where a community or organization we care about is on the cycle. That tells us what kinds of dynamics we can expect and how to understand its challenges and opportunities. So how should we understand where we are in the cycle? For our smaller-scale communities—for example our bioregions, cities, towns, and workplaces—we need to research their histories, their current capacities, and their likely thresholds before formulating our estimates. We have to keep in mind that these communities are affected by what is happening on larger scales, such as in our country and our global community. At these larger scales, we have access to a great deal more data and more scholarly debate about relevant considerations. Thus, our estimates may have a stronger basis, though it is doubtful that we can all arrive at consensus about such topics. Our estimate must be conditional and open to change with new information.

On the global scale, several considerations support the hypothesis that we are in a late conservation phase nearing key thresholds that could lead to severe release.[16] First, as we saw above, we are close enough to planetary boundaries that we live under the threat of undergoing serious release that would cause massive disruption for large swaths of the human population. Though the climate change threshold of approximately two degrees centigrade above preindustrial levels is highly salient, more obscure thresholds associated with biodiversity loss, land-use change, and nitrogen pollution are equally serious threats. Second, on more regional scales, some ecosystems are already entering release, including anoxic zones in the ocean, some commercial fisheries, permafrost systems above the Arctic Circle, and glacier systems around the globe. Many other regional ecosystems seem headed toward release, including African ecosystems structured by elephants, some coral reef systems, and the Amazon rainforest system.[17] Release on regional scales can trigger release at larger scales.

Third, over centuries the Global North has seen a high level of growth. This growth often involved colonial activities that unjustly burdened the Global South, but it also resulted from massive technological changes. Currently the amassing of resources among the wealthy and the strengthening of rules that protect wealth suggest movement into a conservation phase. Our relatively weak international organizations like the United Nations and the International Monetary Fund have not proved sufficiently successful at helping us work together to manage the threat of crossing planetary boundaries.[18] Fourth, international funding to help the Global South develop without using fossil fuels has been woefully inadequate, which places a large portion of the global population in an unjust and untenable position. This is bound to result in more environmental degradation. The failure of institutional processes to prevent the rapid approach to major thresholds is a symptom of being in the late conservation phase near the edge of release.

The evidence for being in a late-stage conservation phase seems even stronger when we look at the scale of the United States. Over the last 150 years, the country has seen tremendous growth in prosperity and capacity to address issues. After becoming an economic powerhouse and contributing to victory in the two twentieth-century world wars, it became a global leader. At some point in the last century, its growth phase shifted toward conservation. Key structures started to become more hierarchical, rigid, and cumbersome. They lock up in private hands considerable resources that are not then available for addressing challenges. Wealth has become concentrated in a smaller set of people. Despite the expanding role of government and rising standards of living, a recent Pew Research Center report indicates that "the public sees an America in decline on many fronts."[19] Increasing political polarization, decreasing trust in governance bodies, and a fragmented media have made it very difficult for the country to address its challenges politically. These trends suggest risks of crossing thresholds that would shift the identity of the country are increasing significantly. Some forms of severe release, such as widespread violence or an authoritarian revolt, are championed by subsets of the population. Even if the US were to manage its conservation phase effectively, its efforts may be swamped by larger-scale failures to address planetary boundaries, or by global economic depression, or increasing global violence.

Of course, it is possible that the US will be able to navigate global disturbances and return to a high growth phase or at least maintain its conservation phase into the foreseeable future. The same could be said for global governance pertaining to planetary boundaries, although our current evidence suggests otherwise. What looks like a release phase may quickly return to a conservation phase that retains the structure and major functions of the pre–release phase. Alternatively, a minor transformation of system elements may lead to renewed growth, as when a company introduces a new popular product line that shifts the focus of the company without changing its fundamental identity. Uncertainties about the application of the adaptive cycle to a given system can be used to stimulate fruitful debate among stakeholders about the dynamics of the system in question. Such debates typically lead to gathering more data about the system and critically examining the values that play a role in system identity characterization. While we might wish for models to be clearer and more certain, we must be content with more provisional understandings that are strongly enough supported to guide action.

If we are struggling to avoid major release, it behooves us to reexamine the cultural assumptions that have brought us here and to consider whether the skillful habits we emphasize are best suited to our situation. Each phase of the adaptive cycle has characteristic kinds of challenges and opportunities, and research is beginning to identify competencies needed to navigate each phase.[20] The entrepreneurial skills emphasized in reorganization and growth phases are less central to the highly structured conservation phase, though still important on small scales. The ability to collaborate across diverse groups is less central to the growth phase and less reinforced in the silos of the conservation phase, but both are crucial for managing wicked problems associated with major thresholds, and necessary for successfully navigating release. As we look at the opportunities that characterize our situation, we should keep in mind the skillful habits that would enable us to make the most of them.

Grasping the Opportunities at the Edge of Release

Our apparent position in the late conservation phase on both global and regional scales highlights two general groups of opportunities: those related to the goal of remaining in the conservation phase and those associated with preparing for and navigating release and reorganization. Both kinds of opportunity are associated with resilience. To maintain the conservation phase, much resilience work involves building reserves and redundancies that will help to prevent an increasingly rigid system from moving across a key threshold. Walker and Salt define resilience as "the capacity of a system to absorb disturbance and reorganize so as to retain essentially the same function, structure and feedbacks—to have the same identity."[21] But sometimes we cannot prevent release and may even try to manage release to achieve some transformation. In such cases, we must understand resilience as "the capacity to successfully navigate ALL stages of the complex adaptive cycle."[22] Each kind of resilience involves distinctive opportunities.

Numerous opportunities lie in identifying system thresholds, communicating with affected populations, and building buffers that keep a community away from problematic thresholds. The Rockefeller Foundation's 100 Resilient Cities initiative is just one example of large communities around the world grasping the opportunity to expand our knowledge about how to strengthen urban resilience. For example, the greater Miami area resilience planning effort wrestled with how to withstand rising sea levels and increasingly severe hurricanes. Sea level along Miami's beaches has risen four inches since 1992, and it is expected to rise another three to seven inches by 2030. Over fifty-three thousand dwellings in the greater Miami area are less than three feet above mean high tide. No threshold has been identified for when such dwellings may be unoccupiable, but hurricanes Andrew (1992), Wilma (2005), and Irma (2017) have already caused tremendous human suffering and billions of dollars in losses. Other stresses affecting resilience, including aging infrastructure, income inequality, and traffic congestion, are also addressed by the fifty actions described in Miami's Resilient 305 strategy. Miami cannot stop sea level rise, but it can restore coastal wetlands, build reef defenses, reduce CO_2 emissions, put in place development restrictions, and plan better for storm surges. These buffers reduce the likelihood of catastrophic impacts from flooding over the medium term, and they create opportunities for a wide range of meaningful work and citizen engagement.

Systems research is crucial for avoiding major thresholds. For example, we will need to invest more in determining where critical thresholds lie and what cost-effective practices will keep us from crossing thresholds. We have learned a good bit about potential economic thresholds beyond which a major recession may lie. Experts at central banks know how to stimulate the economy to keep it away from that threshold, as many nations did during the COVID-19 pandemic. But we have no idea whether there is a threshold for distrust of government beyond which a state becomes effectively ungovernable and hence unable to solve its problems. For many systems, our knowledge of key thresholds is in its infancy.

Technological innovation is a common approach to developing threshold buffers that enable us to reduce our negative impacts on the system. This has been one of our standard approaches to resilience, and it contains a wide array of opportunities, from designing new technologies to developing markets for them and maintaining them. For example, to reduce the likelihood of crossing climate change thresholds, we can improve solar and wind energy technology, enhance battery technology for electric vehicles, and create smart microgrids. The rapid development and production of COVID-19 vaccines have shortened the pandemic and enabled us to approach herd immunity with less death than would have happened otherwise. Here technological development helped us to transform a negative system—the pandemic—by pushing us across the herd immunity threshold. Technological innovation alone won't enable us to fully address wicked problems effectively, but it is likely to be part of the solution.

A third group of opportunities involves developing policies and public/private partnerships that incentivize behavioral changes to help us avoid thresholds. The quickest way to change the behavior of individuals working within a system is to shift policy, hence the battles around such government policymaking as a carbon tax or vehicle fuel economy standards. But policies can also be promoted through nonprofit organizations and businesses and implemented through partnership arrangements. Whether we are concerned about the health of our oceans or the loss of biodiversity, achieving racial justice or reducing income inequality, the opportunity to influence policy is ubiquitous in times of significant change. Take the diversity, equity, and inclusion work that became a crucial part of higher education as a more diverse array of students attended schools. Student advocacy and public protests highlighting racially unjust practices have deepened and broadened this work, stimulating numerous policy reforms. Corporate human resource offices have offered training and adopted policies to embed more inclusive practices throughout their organizations. A great many people have grasped the opportunity to advance this work, but much more needs to be done if we are to avoid the system dysfunction that injustice creates in our society.

A fourth set of opportunities addresses behavioral threats to our communities that we cannot alter through policy. Education, moral suasion, and artistic expression often build support for major shifts in cultural norms. As congressional gridlock, entrenched interests, and insufficient public concern have limited the ability of the United States to adopt powerful national climate policy, schools have played a key role in educating youth about the mechanics and costs of climate change and the actions that must change to mitigate it. Authors and artists have dramatized the impact of climate destabilization. For example, Olafur Eliasson, an Icelandic artist, created Ice Watch, in which he transported huge blocks of ice that had broken off a Greenland ice sheet to London, Paris, and Copenhagen, so people could witness them melting away and contemplate the impact of our collective actions. The emotional impact of dance, theater, music, and poetry are as important as the information from the sciences in motivating us to shift entrenched but dangerous behavior patterns.

Turning to navigating the release and reorganization phases, we find many opportunities that are rarely celebrated because the prospect of release is so unpleasant. Yet it is even more important to grasp the potential upsides of system transformation and the skills that enable people to successfully navigate the transformation process. We cannot know whether we can avoid severe release at a global scale, but we do know that the challenges at this scale are sufficiently great that they will cause many smaller-scale systems to go into release. Most of us are likely to experience the chaos of release close to home. We must hone the skills to make the best of it.

Release provides an opportunity for people with little leadership experience to become emergent leaders and to have a large positive impact on others' lives. When leadership hierarchies from the conservation phase are distrusted and become ineffective, others must take matters into their own hands. Within the resulting chaos, courage, creativity, and equanimity matter. Those who have the ability to focus on what is most important and to move people to action can rise. Those who can improvise action without rules and procedures and those who are calm when things fall apart are best able to marshal the resources to move a group through release. We may immediately think of people who become heroes, like the first responders during 9/11, but often such emergent leadership is more subtle.

Consider the example of Prescott College, whose board of trustees declared bankruptcy in 1975, voted to close the college, and sold off the campus to cover bills. Prescott had been founded eight years before through a Ford Foundation initiative to create a college that would provide the kind of experiential education necessary for the next century. A group of faculty and students refused to let the college go under. They pitched in their own resources and started holding classes in a hotel basement. They focused on the immediate survival of their community. Leaders emerged, and they improvised a form of community governance, while they tried to acquire the legal rights to the Prescott name and the resources to operate a college. The story became national news about "the college that would not die." Eventually, they rebuilt a healthy institution but with a much less hierarchical structure and without many of the frills associated with modern higher education. The faculty of that time recall the anger, the stress, and the pain of the crisis, but they dwell upon the empowerment and the creativity of those days. They stepped up to save a form of education they deeply loved and made a large difference in many lives. We often do not find such opportunities without a good crisis.

The interpersonal opportunities of release have benefits that can last a lifetime. Individuals in releasing systems are highly vulnerable; they must depend on others, which puts a premium on quickly strengthening relationships. Tight bonds are formed through shared struggle and mutual aid. Navigating release also involves gathering resources from new places, often in partnership with outside organizations. Crafting viable partnerships from a position of vulnerability exercises skills useful in many other contexts. A shared crisis is an opportunity to combat the loneliness that many find in an intensely individualist society.

Once a community has stabilized after going through release, it must find a reorganization strategy that promises to lead to a new growth phase. The Prescott community needed to reorganize its governance and physical structures and some of its educational offerings, but it retained its educational vision. Effective reorganization after a profound release requires community members to experiment rapidly with alternative ways to move forward. They must find the resources for the experiments, often shifting directions when experiments do not yield positive results and reframing their vision of success. They must be comfortable with fluid structures. The danger in this phase is that the group cannot converge on a good strategy for regrowth. Eventually funding for experiments will dry up, so it is crucial to choose a clear direction in a timely fashion. The opportunities here include crafting useful experiments, accessing information from outside sources regarding what might work for the community, and crafting powerful narratives for a reframed vision. In some cases of transformation, the general direction of reorganization is clear, especially where some negative state of the community is driving the move to cross a threshold into release and reorganization. But often the direction of reorganization remains obscure; those who are comfortable with action under uncertainty flourish.

Greensburg, Kansas, was hit by an EF5 tornado in May 2007. This rapid release leveled 95 percent of the city and killed twelve people. The rebuilding of the city illustrates some of the opportunities and challenges in reorganization. Like other rural communities, Greensburg had been struggling, but it was not ready to give up after the tornado. The city council chose to see the storm as an opportunity, and they elected to rebuild the city in a way that might attract more growth. True to its name, it became a "green" city. Residents rebuilt all municipal buildings to LEED Platinum standards, mostly with funding from disaster relief. There was a fair amount of skepticism initially about the ideas, especially novel ideas like LED street lighting, rain gardens, and low-flow toilets. The energy efficiency of its schools, medical center, and other municipal buildings now saves the city over $200,000 annually, which helped to convince many. Though city leaders shared a broad vision, they knew they were learning as they went. They tried an experiment of installing municipal wind turbines, but these were too expensive to maintain, so the community shifted to purchase power from a utility-scale wind power producer nearby. It created a new industrial park designed to attract green companies; but this experiment still stands empty. The city received excellent publicity, and many residents stayed and rebuilt their homes. Greensburg achieved a stunning reorganization but has not yet been able to turn this into growth in jobs or population. City leaders have had to reframe their vision and learn to focus on their successes, not the failed experiments.[23] Their courage and innovation have kept this tiny city on the map.

Perhaps the most rewarding opportunities that come with reorganization involve joining with others to move toward a positive vision. The pain of release is past, creativity is welcomed, and the capacity for rapid progress is high. The success of experiments can easily be turned into powerful narratives about forward movement—about the next growth phase. If indeed rapid progress toward sustainability will require significant forms of release, then preparing for heady days when a sustainability transformation kicks into higher gear is an opportunity many can joyously embrace. Imagine different versions of the Greensburg story popping up everywhere. Kim Stanley Robinson's book *The Ministry for the Future* provides a compelling narrative of how this might come about fairly soon.[24] In such a reorganization, people from many different backgrounds could find ways for their skills to contribute to the massive change.

We have seen that uncertainty and socioeconomic turbulence are elevated during this age of climate change. The challenges are daunting, but the opportunities are exciting. Perhaps most daunting is the sheer number, scale, and complexity of our global challenges. Given our recent track record, it is easy to despair of surmounting these challenges. If they are taken individually, we will almost certainly fail to meet many. But sustainability requires us to look for key leverage points that can address many challenges with one move. We must find approaches that enable actions to be unified, easy to communicate, practical for ordinary people, and focused on addressing root causes, rather than symptoms. It is my contention that shifting the skillful habits our culture emphasizes is a promising leverage point.

We can flourish individually now whether or not we approach sustainability. What is required is that we effectively engage with at least some of the opportunities we find. My characterization of the opportunities in the age of climate change has inevitably been schematic and selective. To move beyond broad categories of opportunities would require much more detail about specific challenges and the context in which they are engaged. This overview is sufficient for us to glimpse some of the skillful habits each of us will need to navigate the phases of the adaptive cycle that shape our lives. Recall that these habits include mindsets and emotional capacities, not just behavioral skills. To flourish in this time, our patterns of thought, feeling, and action will need to be reasonably aligned with the challenges and opportunities we face. These patterns—character traits—are not novel. Most of us have them to some extent, but they are underdeveloped. As we will see in the coming chapters, the skillful habits we need to flourish individually will also be crucial for leveraging the sustainability transformation.

COLLABORATING WELL
IN A COMPETITIVE CULTURE

The ability to collaborate effectively with diverse peoples will be especially valuable during the age of climate change, but unfortunately, sophisticated collaboration skills of empathy, hope, and trust are in short supply in the general population. In our highly competitive culture, these skillful habits are often misunderstood, emotionalized, and under-rewarded. Consider the discussion of empathy in Sonia Sotomayor's Supreme Court confirmation hearings. President Barack Obama had said on the campaign trail that he wanted to appoint judges with empathy for marginalized people who most needed the protection of the law. For example, in a 2007 speech at a Planned Parenthood conference, he said, "And we need somebody who's got the heart—the empathy—to recognize what it's like to be a young teenage mom. The empathy to understand what it's like to be poor or African-American or gay or disabled or old—and that's the criteria by which I'll be selecting my judges."[1]

Shortly before nominating Sotomayor, he stated at a press briefing, "I view the quality of empathy, of understanding and identifying with people's hopes and struggles, as an essential ingredient in arriving at just decisions and outcomes."[2] But this view was controversial. Numerous Republican senators vigorously opposed Sotomayor's nomination based on their concerns that empathy was inconsistent with objectivity and that it implied bias in favor of people with similar life experiences. Senator Jeff Sessions put the point forcefully:

> No senator should vote for an individual nominated by any president who believes it is acceptable for a judge to allow their personal

background, gender, prejudices or sympathies to sway their decision in favor of or against parties before the court. In my view such a philosophy is disqualified. Such an approach means that the umpire calling the game is not neutral, but instead feels empowered to favor one team over the other. Call it empathy, call it prejudice, or call it sympathy, but whatever it is, it's not law.[3]

Sotomayor's life experiences and her judicial philosophy seemed to embody Obama's stated criterion. She grew up amid poverty in New York's South Bronx, struggled with diabetes from age six, saw her mother's difficulty financially supporting the family, and witnessed the pain of addiction in friends and relatives. Later, after becoming a federal judge, she argued that her life experiences would and should affect her work as a judge precisely because they gave her an understanding of life circumstances that more privileged people might not have. She famously stated that she hoped a "wise Latina woman with the richness of her experiences would, more often than not, reach a better conclusion."[4] Some senators worried that her emphasis on personal experience might lead to bias and nonrational (i.e., emotional) decision-making. Sotomayor clarified in her testimony, however, that a person from any background could be a good judge, and that empathy did not imply deciding cases in favor of clients like her. She was successful in muting the critique of her views, and she was confirmed on August 6, 2009, on a 68–31 vote. Since Sotomayor exemplifies sophisticated collaborative skills, I will continue to use her as an example of key points.

The issues senators raised about empathy have been applied well beyond the law. They mirror general arguments made by Paul Bloom in his best-selling book, *Against Empathy*.[5] Bloom's arguments focus on the emotional dimensions of empathy—sharing another's feelings—and take such emotional resonance to be a risky basis for much decision-making in comparison with more rational approaches. As we will see, the skills associated with empathy go well beyond sharing emotional states. More sophisticated versions of empathy avoid its alleged deficits.

One reason for our failure to appreciate collaborative skills like empathy is the competitive framing of major institutions like law. In the United States the legal system emphasizes use of competitive skills. We find who is guilty through an adversarial process. We settle many disputes through adversarial lawsuits, and we discover how to interpret the law through a competition between opposing arguments. Empathy appears to have no place here, except perhaps in humanizing a defendant and advocating for lenience. In Europe, however, much legal practice is less adversarial and empathy more respected. Some US legal theorists have championed the role of empathy in legal practice.[6] If we had a greater emphasis on collaboration in our major institutions, we would have more practice in the skills it involves and a better understanding of what expertise in these areas looks like.

Collaboration, like competition, is a family of strategies for achieving goals within a group. A paradigm case of collaboration is when multiple stakeholders voluntarily pool resources to achieve a beneficial outcome that none could achieve alone.[7] The shared resources can include information, influence networks, time, and funding. On the other hand, competitive strategies involve one party exerting sufficient power over others to dictate an outcome, either democratically, economically, or militarily, which the losing parties do not see as acceptable. Whereas competitive processes typically aim at a result where some groups win and others lose, collaborative processes aim at a result where all groups meet enough of their needs and accept the resulting allocation of benefits and burdens as roughly fair.

Many activities involve both collaboration and competition. A common example is team sports, where teammates must collaborate with one another in order to compete effectively against another team. Policymaking is another example. The development of the Affordable Care Act required a great deal of collaboration between insurance companies, drug companies, the medical establishment, and governmental bodies, though it was passed through a largely competitive political process. This point notwithstanding, in group decision-making one family of strategies often dominates. In the Paris Accords process for addressing climate change, collaborative strategies predominated. At a regional level, the Chesapeake Watershed Project was an exemplar of a successful long-term public/private collaboration.[8] By contrast, early attempts to address the severe population decline of the spotted owl in the Northwest forests resulted in a much more competitive process, with loggers and environmentalists each trying to "win" a victory that was unacceptable to the other group.

Although most people acquire at least rudimentary collaborative and competitive skillful habits, they tend to have a preferred approach to group problem-solving. The more we use a cluster of skillful habits, the better we tend to get at them. Once we are good at some skills, we tend to use them more frequently. They become the default, instinctual approach to a set of problems. When most people around us use competitive skills in a social problem-solving context, a social norm is created that reinforces cultivation of more advanced competitive skills. The skill sets associated with the collaborative family of strategies are likely to be less well developed and less salient. The above points are compatible with the observation that the skillful habits we tend to use depend on context. A mostly collaborative person can be quite competitive when playing games, even though that same person is skillfully collaborative when negotiating intergroup conflicts.

The emphasis on competition in US culture is reflected in the nation's version of capitalism, its media, its politics, its dominant approach to leadership, and even its educational system. We celebrate strong leaders who can best opponents. Our media frame public debates in terms of arguments between two opposing sides, with each convinced the other side is benighted and must be stopped.[9] As a result, there is little room for collaborative compromise. Education is largely based on an individual pursuit of learning and grades, rather than growth of the general knowledge of a group. Daniel Markovits argues that the version of meritocracy found among elites in the US fuels hypercompetitive lifestyles, as parents work increasingly long hours to maintain their lifestyles and to maximize their children's opportunities to compete effectively for college placement.[10] Such competition can motivate extra effort, but it is unlikely to cultivate the skillful habits necessary to solve social problems collaboratively. This competitive emphasis has served us fairly well during the growth and conservation phase of the national- and global-scale adaptive cycles, but it is increasingly preventing us from effectively addressing challenges that affect our ability to flourish.

Focusing more on collaborative problem-solving strategies and strengthening the collaborative skillful habits of empathy, hope, and trust will be crucial for flourishing in our times. Empathy helps determine the direction for fruitful collaboration by identifying how individuals in a group can meet enough of one another's needs. With empathy blind spots, we lack the understanding of perspectives and desires that guide potential collaborative partners. Hope provides fuel for a collaborative process. Collaboration around divisive issues is hard; we need to develop the skills of good hope to sustain the pursuit of shared solutions. Without sufficient hope, many people will seek non-collaborative paths for meeting their own needs and ignore the needs of others.

Trust is an important lubricant. Collaboration becomes extremely inefficient if participants have insufficient trust in one another. In such cases, people tend to protect themselves and avoid sharing ideas that increase their vulnerability. This self-policing significantly limits the kind of creative thinking necessary for reframing problems in ways that make mutually beneficial solutions possible. Without the sophisticated skills of trusting well and healing breaches in trust, it is unlikely that we will build the partnerships necessary to effectively tackle wicked problems, and it is likely that the social capital critical for resilience will continue to decline. Before examining these skill clusters in more detail, we should explore the reasons for thinking they are important.

Why Emphasize Collaborative Skills Now?

At least three lines of argument indicate that we should cultivate stronger collaborative skillful habits and shift our emphasis from competition to collaboration. First, as we saw in the last chapter, our position on large-scale adaptive cycles involves increased conflict and the need to rapidly build new relationships, both of which require more advanced collaborative skills. Second, to successfully address global-scale problems that affect daily life, we must be better at collaboration. Third, effectively dealing with local-scale wicked problems involves strong collaborative skills. As we unpack these lines of argument, we will begin to see how specific collaborative skills enhance our flourishing.

1. The relatively stable, hierarchically structured decision-making processes that characterize the conservation phase of the adaptive cycle become less effective in the late conservation phase, where we experience more environmental disruption, social dysfunction, and economic stress. We need better conflict-resolution skills to manage the increasing social conflicts around what to preserve in a waning conservation phase. To avoid crossing dangerous thresholds we will need to significantly change our behavior, which is unlikely to happen without collaborative public processes. To be resilient during such a period, people must rely on diverse groups and work effectively with strangers. New leaders arise; different people tend to be involved in decision-making.[11] For example, the youth climate movement, with leaders like Greta Thunberg and Alexandria Ocasio-Cortez, has shifted the center of power and requires collaboration with a new politically savvy group. We are more likely to flourish in this phase if we have the skillful habits it requires.

Furthermore, after release, groups need to be able to experiment with alternative structures that could support a reorganized community. These groups usually involve novel networks and hence the ability to collaborate with different kinds of people. If release is severe, then most people will become reliant on a very different set of actors. We could see this on a modest scale in New Orleans after hurricane Katrina. Citizens and governments needed to relate effectively with FEMA employees, the Red Cross, church groups providing aid, and neighbors who had access to key resources. The networks that provide basic goods shifted dramatically. We tend to rely on our learned capacity to interact with a small number of important groups in our relatively stable environment. During the release phase of the adaptive cycle, we will need much more robust collaborative skills to grasp the opportunities that lie therein.

2. Most global-scale problems like climate change require collaborative approaches. No country is powerful enough economically or militarily to impose its will on the rest of the world. Even if durable alliances were to create a powerful

enough bloc of countries to determine solutions to global problems, the cost of imposing such solutions on others who do not voluntarily accept them would be high. The lack of strong global institutions and of legitimized global police power would make enforcement of unilateral solutions difficult and costly. Since it seems unlikely that strong global institutions will emerge in the near future, we must continue to pursue the kinds of collaborative agreements sought at Rio, Kyoto, Paris, and Glasgow.

Unfortunately, the potential for reaching effective global agreements is limited if the populations of powerful nations are not supportive of, *and skilled at*, collaborative problem-solving on more local scales. A national strategy of protecting short-term interests and using power to meet needs is likely to appeal to populations with a competitive mindset. Furthermore, international agreements become fragile and harder to fund when nations' support vacillates because of internal political changes. The United States' vacillating support for the Paris Accords and the Iran nuclear deal is a sign of a population divided about the value of collaborative solutions to global problems. Flourishing will be harder in the absence of strong global agreements addressing planetary boundaries.

On a national scale, there is more potential for effectively addressing large-scale issues in a winner-take-all fashion because of stronger governance institutions and police power. Nonetheless, in the US we have recently seen little evidence that this potential is realizable. Indeed, we have vivid reminders of the costs of unilateral problem-solving in terms of congressional gridlock and declining trust in government.[12] Shellenberger and Nordhaus have argued forcefully that the US environmental movement's inability to reduce carbon emissions is a function of its becoming one among many special-interest groups and its failure to ally itself with other progressive social movements to achieve more collaborative solutions to large-scale problems.[13] The Green New Deal appears to be a response to this kind of critique. I acknowledge that sometimes exercise of competitive skills in politics does provide the best results in the long run, but the larger the changes and the less collaborative the population, the more risk there is that competitive processes lead to backlash and dysfunction. We are facing some major changes.

3. Effective collaboration is also crucial for addressing common wicked problems at local scales. Here the link between collaboration and flourishing is salient. Problems like poverty, gun violence, and addiction epidemics have multiple problem definitions, complex causes and effects linking multiple issues, large numbers of stakeholders with competing perspectives, and solutions that involve tradeoffs, many of which cannot be fully anticipated. As a result, it takes complex negotiation among stakeholders to find politically feasible approaches that make advances on the problems. Moreover, if we are to help navigate wicked problems

in an area, each of us needs to understand the systems that create the problem. To integrate the wealth of information required for systems analysis, we need the perspectives of many stakeholders, which typically involves a collaborative process. Thus, a significant portion of a population needs the skills to participate well in public processes. The social learning that occurs in such processes is an important benefit. When such processes are successful, they integrate information about the problem in unanticipated ways. For example, to develop a useful systems analysis of the issues affecting decline of fisheries and water quality in the Chesapeake Bay, a region-wide collaboration reframed the issue as a watershed problem rather than just an issue with the bay itself.[14]

At local scales, the case for collaborative problem-solving has its greatest experiential justification. Wondolleck and Yaffee trace the growing movement toward use of collaborative processes to address natural resource conflicts and describe numerous successes.[15] They argue that collaborative approaches to such conflicts usually lead to better decisions, because more information is available from the range of stakeholders affected by possible solutions. Moreover, such approaches increase the effectiveness of implementation plans because the affected parties tend to support the proposed solution. Lastly, when groups in a region have successfully engaged in collaborative problem-solving, they increase their capacity for dealing with new conflicts, because they better understand opposing parties to the conflict, and they have built trust through personal relations. Collaborative watershed management provides numerous examples of these benefits.[16]

Unfortunately, effective, efficient collaborations are still too rare in the US. Few people have the skills necessary to participate well in them, much less to lead them. The arenas in which we can acquire such skills are uncommon, especially in urban areas, where people can choose to associate only with those who share their views, and institutions tend to be too large for most people to feel empowered. Until we further develop our collaborative skillful habits, we will tend to fall back on competitive approaches to collective-action problems. It is natural to use the skills that we have long practiced, even if they are far from an ideal fit for the context. Without greater capacity to collaborate, we are unlikely to seize the opportunities we will find in this age.

The pressure to double down on competitive strategies is likely to increase over the next thirty years. Resource shortages resulting from climate change and population growth are more likely to motivate self-protection rather than creative cooperation. Recent concerns about immigration seem to reflect a felt need to safeguard self-interest at the expense of people who need resources we possess. Moreover, as trust diminishes, fear of free riders increases, which typically triggers more competitive responses to social problems. In addition, our desire for immediate, decisive action is fostered by both our 24/7 news cycle and by our rapid approach to planetary boundaries. In this context the unilateral action enabled by competitive strategies will appeal more than the murky, time-consuming process of forging durable agreements. To withstand this pressure, we will need stronger cultural norms supporting collaboration and more skills that demonstrate its benefits.

As I have made clear, we should not diminish our competitive skills, just our habitual emphasis on them. We need to exercise strong competitive skills in circumstances that call for them. We will also need to integrate the use of competitive skills with our growing collaborative skills by strengthening what I am metaphorically calling a kind of binocular vision.

Binocular Vision Skills

In binocular vision, each eye provides slightly different data to the brain, and the brain combines those data into a single image. We can see the difference between the images produced by each eye by rapidly shifting between covering one eye and then the other while looking at an object that is partially obscuring an object behind it. The brain must learn the skill of combining the visual data from both eyes, which typically happens when an infant is about four months old. Binocular vision is very useful because it enables direct perception of depth, which allows people to move through the world more efficiently.

Just as our brains can learn to combine different data from our eyes, they can also learn to integrate the information embedded in the schemas associated with contrasting skillful habits. When we engage collaborative strategies, we highlight features of our situation, possible courses of action, risks, and potential benefits. A competitive framing of the situation highlights other versions of these elements. If we have both kinds of skillful habits, we can learn to merge this information into a holistic picture that enables us to determine which strategies are most promising in the context. Sometimes a blended strategy works best. For example, as Russia amassed forces on the Ukraine border, President Biden's national security team needed to assess the range of competitive and collaborative approaches that could be employed to deter aggression. They needed a holistic vision of the factors that affected risks and benefits associated with different possible strategies. After the Russian invasion, they needed to employ strong collaborative skills to mobilize NATO, work with the Ukraine government, and reduce risk of escalation with Russia. At the same time, they needed to make a forceful competitive response to Russia that made that country's war efforts more costly and less likely to succeed. This is the kind of nuanced response that good binocular vision would enable.

Binocular vision does not require that both eyes be equally strong. My right eye is much stronger than my left, and the resulting integration of images is partial. Currently our competitive skills are better developed than our collaborative skills on average, and we habitually frame a wide range of contexts competitively, many of which could also be reasonably framed collaboratively. Whatever binocular vision we have with respect to these strategies is quite limited. By strengthening collaborative skills, we see more clearly the factors influencing their use. We can then habitually integrate these two perspectives, broadening our tool kit for addressing interpersonal situations. In addition, I am arguing that collaborative strategies should become the stronger disposition—the default approach—in most group settings. A version of binocular vision with this collaborative emphasis is most likely to lead to our flourishing in the age of climate change.

Given the tensions between collaborative and competitive perspectives, many will find it counterintuitive to combine them. We must begin by moving between them, seeing the benefits of each, just as we might move between competing interpretations of an ambiguous image like the duck/rabbit illusion. To accomplish this with complex cognitive perspectives requires both flexibility and a fair amount of structural memory. We also need to practice the skill of assessing when a specific purpose requires us to highlight one perspective over the other, while keeping both in mind. Over time, the perspectives tend to merge and appear as a gestalt that has two aspects. Just as a chess master can "see" what will result from a move without rehearsing each intervening step, we can learn to approach any situation seeing how it can be characterized in both collaborative and competitive terms even if we do not consciously work out the details of the characterizations.

Developing binocular vision with respect to contrasting clusters of skillful habits will be a theme throughout the book. We need to develop better binocular vision for humility and conviction, for frugality and abundance, and for systems thinking and individualistic perspectives. Though in each case I am arguing that we need to shift the emphasis and strengthen the underdeveloped skill set, I do not want to lose the contrasting skill sets, thereby creating other problems. The integration of contrasting skillful habits should mute the concern that my position involves taking sides in a culture war. It may be popular to assign contrasting skill sets to opposing liberal and conservative camps, but this practice often obscures the roots of these skill sets in traditions invoked by both camps.

We are now in a position to explore the three clusters of collaborative skills associated with empathy, hope, and trust. In box 1, I identify specific skills associated with expertise in these clusters. These are not the only collaborative skills we must cultivate, but without these, we are unlikely to make headway in building our collaborative capacities.

Box 1. Collaborative skills

Empathy skills

- Understanding and caring about others' thoughts and feelings
- Engaging and detaching from emotional empathy
- Recognizing and expanding the limits of our empathy

Hope skills

- Regulating our agency
- Fostering hopeful communities
- Assessing hope's rationality

Trust skills

- Extending trust widely and wisely
- Building trust in groups
- Repairing trust after a breach

Empathy and Collaboration

We return now to empathy, not as a mere emotion, but as a complex set of skills that are essential to effective collaboration. At its core, empathy is an ability to understand the perspectives of others in all their rich significance. Without empathy, we have a hard time jointly designing a solution to a group problem that meets enough of others' needs to secure long-term support.

One of Sonia Sotomayor's stories illustrates nicely the role of empathy in collaboration. When she was a student at Princeton University, she was a member of Acción Puertorriqueña and Princeton's Third World Center, which were advocating for a more diverse faculty and staff. Many of her peers wanted to take an adversarial approach with administrators, but she argued for a more collaborative approach. As she reflects on this, she says, "My strengths were reasoning, crafting compromises, finding the good and the good faith on both sides of an argument, and using that to build a bridge. . . . A respectful dialogue with one's opponent almost invariably goes further than a harangue outside his or her window. If you want to change someone's mind, you must first understand what need shapes his or her opinion. To prevail, you must first listen."[17] The groups used her skills and did succeed in motivating Princeton to hire its first Hispanic administrator, an assistant dean of student life, whose job was in part to support minority students. The result was not all they wanted, but it was a success on which they could build.

In the above passage Sotomayor is primarily focused on the skills of cognitive empathy—the ability to know what is going on in another's mind. This is typically distinguished from emotional empathy—the ability to share the emotion another is experiencing. And both of these are distinct from concern about what another is experiencing. Thus, Heather Battaly describes empathy as "knowing, sharing and caring."[18] It involves the ability to know what others are thinking and feeling, to share some of those feelings, and to care about those feelings. These elements of empathy may appear separately or together. Researchers often focus on one of these elements to the exclusion of others.

The foundational empathy skills involve "reading" others' thoughts and feelings. Most children acquire rudimentary skills in this area by the age of four. Some people become highly adept at reading others. An excellent poker player will recognize "tells" that signal when someone is likely to be bluffing. A good speaker must be able to read a crowd, and a good business leader must be able to sense what stories will motivate employees. For people on the autism spectrum who lack such skills, social interaction can be challenging. At first glance, emotional empathy may seem less like a skill than an inclination to feel what another is feeling. The same might be said about caring about others' feelings. But skillful sharing and caring involve learning to exhibit these dispositions in the right way at the right time, and to communicate them appropriately.

A more detailed account of empathy skills emerges from an examination of the reasons why objections to empathetic judges we considered above ultimately fail. Kathryn Abrams outlines the three primary objections as follows:

1. Empathy biases us toward groups of people like us because we tend to empathize more with those like us. Thus, it reduces impartiality.
2. It compromises rational decision-making, because it is grounded in emotion.
3. It renders us less objective, because it involves situating ourselves within a group and a context.[19]

To be sure, we would not want judges who are routinely partial, irrational, and subjective in their rulings, nor would we want this to characterize our everyday decision-making. As Paul Bloom and Jesse Prinz argue, such attributes are poor guides to ethical decision-making.[20] But I have suggested these attributes apply only to those with underdeveloped empathy skills.

Let's start with the bias concern. A person who excels at empathy recognizes the limitations of her empathetic capacity. She can detect her blind spots and biases. She is good at listening and imagining the viewpoints/experiences of distant others and out-groups, including those with whom she disagrees. She has learned how to deepen and broaden her empathetic responses. It is not hard to understand what a close family member is feeling and to share that feeling, but it may take a lifetime to become good at understanding a vast variety of peoples, to be disposed to share in their joys and sorrows, to recognize where such capacities are limited, and to know how to expand such capacities when the situation calls for it. A number of studies suggest that people do tend to empathize more with those who are like them. However, taking this as evidence that empathy is a poor guide to decisions is like arguing that reasoning is a poor guide to decisions because studies show people commonly reason fallaciously. The problem is not with reasoning or empathy, it is with underdeveloped use of the tools.

The second objection presumes a dichotomy between emotion and reason, according to which reason provides a balanced and systematic approach to decision-making and emotion involves nonrational distortion, bias, and even pure fantasy. This dichotomy has been widely criticized.[21] Rather than seeing emotion and reason as competing approaches to decision-making, we should view them as interwoven into the fabric of human understanding and jointly involved in much decision-making. For example, emotions are often justified by the reasons for beliefs that generate the emotion. I might be angry at my brother because he has not repaid a debt. My justification for the anger involves my reasons for believing he has not repaid a debt and for thinking that he should have done so. Moreover, we often use the insights of emotion to generate the premises from which good reasoning proceeds; my joy at a surprise visit from an old friend may lead me to reason that I should do more to nurture my friendships. Once we rid ourselves of the reason/emotion dichotomy, we can develop more sophisticated views of empathy.

Collaboration involves developing skills of rationally negotiating our own and others' emotions. Our emotions often convey vital information; they reveal meaning and enrich understanding, and underlie much motivation. Emotions need not compromise rationality if they are wisely managed. The skills of good empathy include not only the ability to share deeply in others' perspectives and feelings, and to communicate this sharing effectively, but also to *detach* from this shared feeling in order to evaluate what to do in the context. Imagine two parents whose child is involved in an escalating fight with another child. They need to emotionally appreciate their child's perspective but also do the same with the other child. Then they may be able to find a way to de-escalate the anger and reach a solution that each child can accept. The parents must share and care about the feelings of both children. Emotional empathy here does not imply agreement; rather it adds to the depth of understanding of how the children frame the situation and what motivates them. The trick to integrating emotional and cognitive empathy into good collaborative decision-making processes is to hone our skills of engaging fully with others' emotions and then detaching enough to fully consider other relevant information. The skill of engaging and detaching from emotional empathy is crucial for good leaders, and also for judges. In a collaborative setting this depth of understanding is very helpful in developing ideas for workable solutions to a problem. It makes collaboration more effective.

The third objection also targets the emotionally laden perspectives associated with empathy. An emotion situates us somewhere in the world where having it makes sense. If I feel anger, I am in a position of having been mistreated. But the objection then states that situatedness reduces our objectivity. On this view, objectivity can only be complete if we approximate a point of view independent of all situatedness. Thus, the objection claims that judges should try to reason from the law alone, and not from their particular situations or those with whom they empathize. But can we really approximate such objectivity? Hilary Putnam describes this view of objectivity as trying to achieve a "God's eye point of view," which is impossible for us.[22] We must all engage with each other from our situated points of view, but empathy enables us to adopt alternative points of view, which may temper our own perspectives. Sotomayor quotes Martha Minow in defending a similar view of objectivity: "There is no objective stance, but only a series of perspectives—no neutrality, no escape from choice in judging."[23] As Kathryn Abrams notes, "The judge who is truly dangerous . . . is not the judge who recognizes her situatedness and its contributions to the inevitable choices she faces in judging. It is the judge who does not understand she is situated at all. . . . A person who understands her individual perspective in this way has no incentive to look for anything beyond it when she considers a case."[24]

The skill suggested by this response to the objection involves recognizing how our—and others'—emotionally colored points of view shape thinking and action and guarding against viewing any perspective as the uniquely objective view. This skill is also an element of humility, which is the subject of the next chapter. This skill enables us to better see the limitations of multiple points of view advanced in a collaborative setting and to creatively adjust proposed goals in ways that maximize broad support.

Once we become aware of some component skills of sophisticated empathy, we see why the objections to having an empathetic judge miss the mark. We also see that most of us fall short of excellent empathy. We can become better at reading others' thoughts and feelings, at recognizing and expanding the limits of our empathy, engaging and detaching from shared emotions in decision-making, and challenging the points of view we bring to a context without becoming skeptics. I have illustrated some benefits of such skills in collaborative process, but they are also highly beneficial in building new relationships quickly. This is particularly important when we must navigate release.

In becoming better empathizers, we realize how much more morally complex the world is; we become much more aware of the tradeoffs we are making; and we see that the challenge of managing conflict is ever present. In effect, having strong empathetic skills makes us highly attuned to the challenges of finding just win/win collaborative solutions to problems. To rise to these challenges, we need to build our capacity to *hope* for sustainable collaborative solutions to our problems. Hope provides the motivational energy to persist even when the odds of success are low.

Hope and Collaboration

As with empathy, many people underestimate the significance of hope because their understanding of it is underdeveloped.[25] We often use the word "hope" to express a mere wish that something occurs. For example, I may "hope" that you are having a nice summer or that your exams went well. This sort of hope is usually explained as involving a desire that something happen and a belief that satisfying the desire is possible, but not certain.[26] This view of hope involves no evident skill or action-orientation. Such hope can easily be written off as wishful thinking, like the "hope" that other people will solve the climate issues without requiring any sacrifice from us. By contrast, the kind of hope that signals a strong sense of agency and involves key skills regarding regulating agency is a set of skillful habits one can cultivate. These make up what Victoria McGeer calls "the art of good hope."[27]

The importance of hope is nowhere more salient than in our response to climate change. If we think that technological progress will adequately address climate change, then we do not need to have hope regarding this goal. We can simply be optimistic. However, as we saw in chapter 1, this optimism is unlikely to be warranted. As Paul Hawken puts the point, "When asked if I am pessimistic or optimistic about the future, my answer is always the same: If you look at the science about what is happening on earth and aren't pessimistic, you don't understand the data."[28] Byron Williston has a similarly grim outlook.[29] The probability that we will avoid significant climate disruption change seems very low in light of our history of inaction, the difficulty of reaching just, sustainable agreements to decarbonize the economy, and the long lifespan of atmospheric carbon. A natural response to such pessimism is self-deception and/or despair. We either avoid looking at the facts or we give up on solutions and try to maximize the enjoyment of our current lifestyle before this becomes impossible. As some might say, we party while Rome burns.

Nevertheless, Williston argues that effective responses to climate change require developing the virtue of hope: a full emotional and behavioral engagement with the pursuit of the hoped-for goal despite its low probability of success. His justification is pragmatic. If we despair, we lose our capacity to act. As a result, we have a moral obligation to develop our capacity for hopeful action. Here, hope is action-oriented; it supplies the motivational energy to pursue long-shot goals. Without applying hope skills to climate change, we will all be worse off because we will further reduce our odds of rapid decarbonization of the economy.

Paul Hawken follows his point about pessimism with the observation that "if you meet the people who are working to restore this earth and the lives of the poor, and you aren't optimistic, you haven't got a pulse."[30] His use of optimism seems very close to the kind of hope I have in mind, since he takes it to be compatible with a pessimism based on the facts. This is different from the more common view that having optimism about a goal involves the belief that one has good evidence that one will reach the goal. Given this common meaning of "optimism," we should *not* be optimistic about rapidly decarbonizing the economy, but we can be hopeful. Engaging with those who are working hard to make that long-shot goal a reality helps to fuel our own hope.

The skillful habit of hope involves finding new challenges stimulating and redoubling efforts when reaching a goal looks less probable. A sports team that often comes from behind to win a game seems to have such capacities, and its coach or team captain might have the skills to stimulate hope in team members. As C. R. Snyder notes, "The advantages of elevated hope are many. Higher as compared with lower hope people have a greater number of goals, have more difficult goals, have success at achieving their goals, perceive their goals as challenges, have greater happiness and less distress, have superior coping skills, recover better from physical injury, and report less burnout at work, to name but a few advantages."[31]

One can break the skillful habits associated with hope into three interrelated categories: skills focused on self-regulation of one's own agency, skills involved in fostering hope within a community, and skills required to assess the reasonable application of hope in a context. The main skills involved in regulating our own agency include focusing on goals and past successes, rather than on constraints and past failures, rehearsing hopeful cognitions to boost motivation, and developing multiple creative pathways for achieving a goal. They also include social skills that assist one in accessing a community that helps to bolster one's agency.

McGeer draws a strong connection between self-regulation of agency and engaging in a hopeful community, because the development of hope in early life depends on the scaffolding provided by others.[32] Young children are relatively helpless; they typically need the encouragement of caregivers to learn how to overcome frustrated desires and to persevere in pursuit of some new goal. The parent who repeatedly says "you can do it" after the child has made several failed attempts is in part fostering the development of hope. A parent may also suggest taking different approaches to the problem, or recommend taking some time out before returning with new energy to the goal. Good parenting provides the child with a range of skills that can be drawn on to accomplish increasingly challenging goals; but as the child becomes an adult, parental scaffolding of hope needs to be replaced by peer scaffolding. We can develop friendships that help us to rise above our challenges and to deepen our agency. Knowing how and when to access such peer relationships to strengthen our hope is a key part of regulating agency. Those who fail to develop such friendships are more likely to fall into despair.

Martin Luther King Jr. is an exemplar of the skillful habits of hope, especially communicating hope. This skill requires knowing your audience—having appropriate empathy, so that you can use language and cautionary tales that will strengthen motivation. It involves being able to describe the depressing facts of a situation without diminishing hope and articulating the dynamics of hope and despair using familiar examples from life experience. His famous "I Have a Dream" speech gives voice both to the long struggle of Black people to achieve justice in the United States and to a renewed determination to make "all men are created equal" a reality. As he urges his audience not to give up, he says "This is our hope. This is the faith that I go back to the South with. With this faith, we will be able to hew out of the mountain of despair, a stone of hope."[33] The speech is a powerful rallying cry for a people to continue to work toward the goal of genuine justice; it is a masterpiece of hope communication.

Anyone who has been a leader of an organization or community knows how important it is to be able to foster hope. As a provost in a small college, I often tried to tap faculty skills of hope when internal conflicts threatened to prevent us from overcoming our obstacles and achieving some important objective. Often this involved creating opportunities for peer-to-peer problem-solving that would renew the hope that together we could achieve something. No great orator, I found that my best skill for fostering hope involved a kind of coaching that helped build confidence that together we could succeed.

The above skills will often fail to generate good hope unless one can also assess the limits of reasonable hope in a context. Appropriate assessment skills increase the chances that one avoids wishful thinking and unreasonable optimism. As a teacher, I sometimes had students who had great motivation for learning and good work habits but whose goals surpassed their abilities by a significant margin. They hoped for A's when B's would be a significant achievement. Pushing oneself hard to earn an A is often worthy, but not when falling short risks despair. A cancer patient may hope for a full cure, but when she learns that the cancer is terminal, the skills of good hope require her to adapt her goals so she hopes for something she can reasonably work hard to achieve—perhaps a good death. There is no rational calculus for assessing the appropriateness of hope. That involves weighing the constraints on achieving a goal, the importance of the goal, the motivational limitations of those who must act, and the potential for unforeseen roadblocks. In the case of climate change, it is the tremendous moral importance of rapidly decarbonizing our economy that warrants hope, even when the constraints seem to make the probability of success very low.

Sometimes the severity of the constraints on achieving a goal makes it very hard to grasp what achieving the goal might look like. In such cases, we must frame the object of hope in a very general way. Jonathan Lear called this "radical" hope and attributed that trait to Plenty Coup, the chief of the Crow, who urged his tribe to adapt to the onslaught of white settlers rather than to fight them.[34] His hope was that his people could retain some of their identity, culture, and sense of agency by agreeing to live on a reservation that preserved a portion of their ancestral lands. He did not know how this could be possible or what it would look like, but he still managed to persuade the Crow to pursue these goals, rather than to fight to the death. Both Thompson and Williston argue that we must cultivate such radical hope that we avoid climate catastrophe.[35] We cannot fully understand what achieving that goal will look like and what sacrifices it will require. The skills associated with radical hope are necessary for increasing our own motivation to make sufficient changes in our own lives, and also to support the collaborative processes necessary on a global scale to avert the crisis.

The skillful habits of good hope apply to action on all scales, but to strengthen our skills we must practice them on the local scale, that is, in our personal lives and in our local organizations. Not long ago, I was asked to moderate a forum for a new organization focused on clean air in the Glens Falls, New York, area. After an excellent talk by an expert on pollution and public health and a short panel discussion with local officials, audience members began to express their frustration with local inaction regarding a large trash incinerator and several other polluting businesses. Local officials became defensive, and a cynical tone started to dominate the proceedings. Instead of raising public awareness and building support for this new organization, the event threatened to leave people angry and disempowered. It was an opportunity for several of us to practice skills of good hope. We described the dynamics we saw in the group, explicitly empathized with the full range of stakeholders, identified actions the group could take that might be productive, highlighted hopeful voices, and obtained clear support for some action steps from local officials. Some anger and cynicism remained, but most people agreed to come back for further meetings to see what we could achieve together. This small but common kind of situation suggests how regularly we can find opportunities to practice the skills of hope. In interpersonal relations, in group projects for a class, and in interactions with friends and relatives, we inevitably make tacit choices about whether to foster hope.

Trust as a Collaborative Skill

Without hope, participants may check out of a difficult collaborative process. Without trust, they are likely to create premature stalemates where no one is willing to make the first move to break through entrenched conflict. In low trust contexts, most people try to protect their interests and avoid being manipulated. In extreme cases, such as intractable international conflicts, a series of trust violations between the parties makes each side fear that any trust will be taken advantage of. No one is willing to look weak by suggesting compromise. In most polarized political decision-making processes, lack of trust leads to seeing disagreement as a fight to be won rather than a compromise to be fostered. Those with whom one disagrees become enemies to be defeated.

Some amount of trust is required to start the work of creative solution development that will garner support from most stakeholders. Thus, many collaborative processes begin with trust-building exercises in which relationships between participants begin to be built. The more trust that can be developed, the more vulnerable and experimental people can be, which is essential to effective and efficient collaboration. Outside of explicit collaborative settings, trust is crucial for smooth interpersonal interaction.[36] One can ask directions without fear of being fooled, and receive change in a transaction without counting it. These small efficiencies magnify as the stakes increase; in business, organizational partnerships, and marital relations such efficiencies are part of what make relationships work well.

How should we understand trust, given its evident value? My answer owes a great deal to Solomon and Flores, who distinguish "authentic trust"—a *choice* that one makes in full view of the potential for trust violations—from the feeling of trusting.[37] Authentic trust requires one to extend trust not through a calculation regarding trustworthiness but rather as a way of valuing the potential relationship and increasing the probability that the other will live up to the trust. For example, a parent may continue to extend trust toward a teenager when evidence of trustworthiness is ambiguous, with the hope that the teen will live up to the trust. Such trust is contextual and conditional. It depends on the specific situation, and it includes an awareness of the limitations on how far trust should be extended. In this sense it is the opposite of a naive trust that is unreflective and automatic.[38]

Like empathy and hope, trust is often emotionalized. It is thought to be a feeling of safety we have either because of our innocence or our experience of someone's trustworthiness. Both of these are problematic. Innocent trust may be appropriate for children, who may reasonably take for granted the trustworthiness of caregivers. But such trust does not acknowledge the possibility of betrayal, so it is fragile if it becomes the model for adult trust. Once such trust is violated it cannot be regained. The innocence is lost. On the other hand, if trust is just a feeling we have once trustworthiness is demonstrated, then we risk not gaining the benefits of trusting those who have not yet had a chance to demonstrate trustworthiness and those with whom we might repair a trust breach. This latter view of trust might make sense in a highly competitive society where we might presume that people we do not know are likely to take advantage of us to maximize their own self-interest. It is ill-suited, however, to times where we need to foster collaboration with those we do not know well and where we need to build new relationships quickly.

Trust is better viewed as a set of skillful habits that include trusting wisely, building trust, and repairing trust breaches. Trusting wisely involves making good judgments about the limits of reasonable trust in a context. This skill requires cognitive empathy—an understanding of the motivations of others; it also involves an assessment of their competence in acting on those motivations. Some trust breaches occur because people entrusted with a project are not competent to carry it out; others result from people having insufficient motivation to carry it out. Another aspect of trusting wisely is assessing the impacts of potential actions on the dialectic of trust. This involves imagining what can be achieved through greater trust, and understanding how to build trust incrementally through a series of small successes. A third aspect involves retaining a vivid awareness of the risks of trust while extending trust. It is easier to practice such skills in the context of close personal relationships, where the importance of maintaining the relationship is often highly salient. But trust is also valuable when it extends outside our family and friends into a broader set of acquaintances and strangers. The use of trusting skills outside of our established social networks builds bridging social capital, which is especially important during periods of release and reorganization, when one must collaborate in new ways with new stakeholders.

Solomon and Flores emphasize that trusting wisely is not reducible to a complex cost-benefit analysis of risks of trust regarding an action in a context. My judgments that I can trust my mechanic to repair my car appropriately but not to keep a secret need not be based on cost/benefit analyses. The calculative approach to trusting wisely seems to presume that we can assign probabilities to the benefits that come with trust, but many of these are unpredictable. Also, the calculative view usually suggests a self-interested focus, whereas trust as a skillful habit is focused on the value of the relationship.

McGeer amplifies this critique of the calculative view by drawing an explicit link between hope and trust.[39] She argues that trust can be rational despite lack of evidence of trustworthiness because we can be pragmatically justified in feeling hopeful about the person's dispositions to act in our interests. We can also be hopeful about the role that our trust plays in strengthening those dispositions. We can be hopeful that trust will beget trustworthiness. Trust is a voluntary acceptance of vulnerability to another—a kind of believing in the other's agency—that we reasonably hope will be appropriately honored. Thus, a trusting person must also be good at hope, *and* not too risk averse.

Some might object at this point that extending trust to acquaintances and strangers is not rational in the context of a fifty-year decline in interpersonal trust. Putnam and Garrett found that interpersonal trust peaked in the mid-1960s when approximately 55 percent of survey respondents said most people could be trusted; by 2014 the figure was 31 percent.[40] The average interpersonal trust of generational cohorts seems to stay fairly stable, but younger cohorts have markedly less trust than older cohorts. Other surveys over the same period find similar decreases in trust. Gallup has assessed how much confidence Americans have in institutions since 1979. In this context, trust and confidence are roughly equivalent. The percentage of people saying they have a great deal of confidence, or quite a lot, in public schools has dropped in the last forty years from 53 to 29 percent; in banks the drop is from 60 to 30 percent; in organized religion 65 to 36 percent; and in Congress from 34 to 11 percent. The only major institution to see an increase in confidence over that period is the military.[41]

This robust decline in trust does not justify withholding trust where we lack evidence of trustworthiness. This would contribute to a spiral of suspicion. If others believe our trust in them is low, then they must protect themselves by not trusting us. Instead of people starting from a default position of trust, distrust becomes the default. But if we react to the decline in trust by refusing to trust acquaintances and strangers, we become isolated, bound by our limited friend groups, and increasingly cynical. Or worse, we hypocritically imply that we trust others when we really do not. Solomon and Flores suggest that this kind of cordial hypocrisy deeply compromises the vulnerability necessary for efficient teamwork within organizations.[42] Nevertheless, we do not need to acquiesce to this default of distrust. In our actions, feelings, and speech, we can rebuild trust, starting in local communities.

The skills of building trust in groups involve being trustworthy, effectively communicating our trust, and owning the ideal of creating a community of trust. Where we are leaders, such skills also include delegating authority, showing that we trust others to do good work; micromanaging erodes trust. Perhaps most importantly, trust building involves using a principle of charity in interpreting the behavior of others. We should interpret actions and speech in ways that put them in the best light—that make them understandable and reasonable—unless there is clear evidence to the contrary. The skills of empathy are deeply intertwined with those of charitable interpretation.

Building trust also involves critiquing untrustworthy behavior in others, while humbly recognizing that few of us can avoid it entirely. If people feel that others get away with breaching trust, they may do the same when it would serve their self-interest. It is easier to erode trust in such ways than to rebuild it afterward. We must be careful here, though. Sometimes we find ourselves in situations where all choices appear untrustworthy. For instance, in any genuine moral dilemma, each action is wrong is some respects. Each action appears to breach some reasonable trust. Only when we understand the entire context of an action, including the choices that are available, can we confidently criticize the untrustworthiness of others.

In addition, developing and sharing a reservoir of trust-building success stories is helpful in breaking cycles of distrust and cynicism. The stories generated by the nonprofit Braver Angels are particularly powerful examples. They show how in the course of a weekend ordinary citizens can break through negative stereotypes associated with Democrats and Republicans and come to trust those with whom they deeply disagree. A great deal of good work is being done to rebuild trust nationally in response to political polarization. Many other specific skills for community-building enhance trust, including conflict resolution, stakeholder engagement processes, and social marketing.

The final cluster of trust skills is involved in healing trust breaches. The skillful habits of trust repair differ from trust building because they focus on addressing a situation where an individual or group has wronged another. A few years ago I taught a course on building trust, and my biggest surprise was the majority view that when trust has been broken, it cannot be repaired. Students' initial response to a significant trust breach was to abandon the relationship and seek elsewhere for someone to trust, even if the violator had been a close friend or lover. Many were not explicitly aware that practices they used when they committed a wrong, such as making excuses or apologies, were actually skills they could improve. Related skills include explaining why a violation occurred, acknowledging the harms it caused, and offering reparations where appropriate.

In a review essay on empirical research regarding trust repair, Lewicki and Brinsfield describe the effectiveness of different repair strategies for different kinds of breaches.[43] For example, apologies are more effective for breaches that result in failures of competence than for actions that appear to come from lack of integrity, presumably because one is likely to be motivated to improve one's competence, but flaws in integrity are less likely to be corrected. Silence is sometimes more effective in the latter cases. Apologies that express genuine repentance and assurance that the violation will not happen again, along with the request to be forgiven, are more likely to be effective than those that leave out such elements. Much depends on the context of the violation, however. Breaches early in a relationship are harder to recover from than those that occur after the relation is well established. Reframing the violation can be important for long-term strategies of trust repair. In a collaborative setting, the violator often must demonstrate a commitment to changing behavior in order for an apology to heal a rift. Genuine forgiving on the part of those harmed will put closure on a breach, even if it is not forgotten. The skill of forgiving well is itself an important part of the trust repair. Trust repair at an organizational level involves similar skills.[44]

The above catalog of trust skills, most of which are familiar, should put to bed the idea that trust is just a feeling. Their familiarity does not mean they are generally well developed, however, as is evident from our ambient level of distrust. Most of us can improve our capacity to foster trust.

Cultivating Collaborative Skills

Fortunately, collaborative skills can be strengthened at any time. The occasions for practicing them and reflecting on the results are ubiquitous. The first steps are to consciously choose to become better at collaborating and, where appropriate, to make such practices the default approach to problem-solving. Sotomayor's memoir provides not only a wealth of examples of such skills at work but also a useful exploration of how mentors and role models can help us cultivate stronger collaborative skills. She tells numerous stories about how she intentionally developed relationships with individuals who could help her strengthen her skillful habits and who had character traits she wished to emulate.

Sotomayor learned very early in life how to find a mentor. In grade school she struggled. Her fifth-grade teacher, Ms. Reilly, put gold stars on the board next to the name of any student who did something really well. Sotomayor wanted those gold stars, but she did not know how to get them, so she asked one of the "smart kids" how to study. This sounds obvious, but it is fairly rare. Donna Renella did teach her how to study, but more importantly, she taught her that often just asking for help can create a mentoring relationship. As Sotomayor puts the point:

"Don't be shy about making a teacher out of any willing party who knows what she or he is doing."[45]

Later, an older student, Ken Moy, who coached the Forensics Club, became both a mentor and a role model.[46] The difference is subtle but important: a role model embodies skillful habits we want to learn but may not know or care whether we learn them. A mentor may not be a strong role model for the traits we are seeking to develop but may know how to help us progress. The mentor's primary tools are linguistic; they include interpreting a situation, providing advice, critiquing performance, and giving praise. A role model's primary tool is behavior. Of course, one may be both a role model and a mentor, as Ken Moy was for Sotomayor. He mentored her regarding oral debate, and he served as a role model for analytical reasoning. He also helped her through the process of applying to a top college.

Later, when Sotomayor was at Yale Law School, she describes José Cabranes, Yale's general counsel, as her first "genuine" mentor. By this she meant a highly experienced professional who took her under his wing and advised her throughout her career. At his invitation, she went to work for him after her first year at Yale and closely observed his interactions with others. She talks of his skill of engaging "with warmth and depth with whoever he encountered" and his ability to "maneuver with equal skill and self-assurance, a kind of courtly grace, in the most rarefied corridors of power."[47] She notes that having a close role model like José Cabranes convinced her that she too could learn to do what he did. At Yale, she honed her skills of judging how and when to collaborate, and Cabranes played an important role in this, as did later mentors she found when she was a prosecutor and when she was in private practice. At each step in her career she developed relationships with people who could teach her how to fine-tune the skills she already had, and to add others in order to succeed in the new context.

In my conversations with peers, I have been surprised at how few people have deliberately tried to find mentors. Many could name a few people who had served as mentors, but few had made it a project to find strong mentors and to build the relationships that would maximize the learning from them. I confess that I have not done that either. I have had some wonderful mentors, but it was more luck and their gracious efforts than my doing. I could have learned a good deal more had I been more deliberate.

Finding mentors is not so hard if instead of looking for a perfect role model we look for insight amid imperfection. We have to know what we want to learn and be willing to watch and listen. Some of my best mentors have been friends, colleagues, or bosses who would not have said they were mentors. They were just open to telling or showing me how they do something. Early in my teaching career I struggled with calling out students who were not performing up to their capacity, worrying that I would breach an already weak relationship in the process. I watched how my colleague Dick Prust could with jovial good humor and evident care call out a student, create a moment of connection in the process, and build trust while also remotivating the student. I could not duplicate his approach, but I developed my own version, which has become a standard tool in my collaborative skills toolbox. This is mentoring writ small, to be sure; it is far from the genuine mentoring relationship Sotomayor had with José Cabranes. It is where we must start, though.

I will discuss other approaches to cultivating broad skillful habits in later chapters. We may achieve a lot through our own initiative and through the support of those we know, but to scale up the acquisition of collaborative traits, we must also look to major institutions, which are inevitably shaped by the dominant competitive orientation of our culture. Education and business are promising candidates. Consider just one data point. Many employers want schools to do more to cultivate teamwork skills. Roughly every three years the Association of American Colleges and Universities commissions an employer survey to assess employers' perceptions of how well-prepared recent graduates are to meet employers' needs. The 2018 survey was consistent with prior surveys in showing a significant gap between graduates' ability to work effectively in teams and the needs of employers in this area.[48] Of the over five hundred executives surveyed, 77 percent said being able to work effectively in teams was very important, and only 42 percent said graduates were well prepared in these skills. In chapter 7 we will see how education is responding to such needs.

I close this chapter with a final story from Sonia Sotomayor's memoir *My Beloved World* that illustrates how cultivating binocular vision is a lifelong process and underscores the role of vulnerability in relationship-building. Recall that Sotomayor has type one diabetes. She had learned how to administer insulin shots herself at age seven. This was one manifestation of her developing self-reliance. Through school, college, and early steps in her career, Sotomayor did not let people know she had diabetes. In general, she rarely opened up about her many struggles. As she says, "Many times I felt there was a wide moat separating me from the rest of the world, in spite of my being, by all accounts, a great listener to all of my friends. . . . Like my mother, I would suspend judgment, feel their pain, perhaps even point out a fact they had overlooked. . . . The only trick I could not manage was to ask the same of them."[49]

While she was working with a private law firm, Sotomayor was invited by a client to a wedding in Venice. She went to her hotel room after the flight to rest, and fell into a diabetic coma. When she did not appear at the wedding, two friends thought something might be wrong. They went back to the hotel and threatened to break down the door when the manager would not let them in. They found her passed out and quickly rushed her to a hospital, saving her life. Sotomayor had become adept at estimating her blood sugar levels based on what she had been eating and managing five to six insulin shots per day, but the system was not foolproof. She realized that she had been extremely lucky to have friends in Venice who had known about her diabetes. She could not count on that luck; she needed to share more about her challenges. She says, "Learning to be open about my illness was a first step, and it taught me how admitting your vulnerabilities can bring people closer. Friends want to help and it's important to know how to accept help graciously."[50] She also learned that many of her difficulties relating to her mother came from her mother's difficulty in being similarly vulnerable. The growing empathy for her mother, and the mutual sharing that it engendered, went a long way toward strengthening that relationship.

For me, this story is layered with insights about the development of skillful habits across a lifetime. We often find that the strengths that enable us to flourish early in our lives carry with them weaknesses that we must address later. Sotomayor's self-reliance was a tremendous asset, but it came with costs. She then had to work on those weaknesses over many years to balance appropriately the skills of self-reliance with the skills of relationship-building through sharing vulnerability. Altering our assessment of where to find a balance between contrasting skills involves sharpening our capacity to make good judgments. The process of developing the capacity for judgment is lifelong. We balance and rebalance, and thereby refine our core skillful habits—our character. We tend to think that character traits are set early in life, but we have a great deal of freedom to change them, even late in life. Sharing our vulnerabilities is also a sign of humility, which is the next cluster of skillful habits we will examine.

RECOVERING HUMILITY AND SOFTENING CONVICTION

The US cultural emphasis on competition is deeply intertwined with our craving for conviction, especially in our leaders. We gravitate toward leaders who are highly assertive, who express no doubts about their actions, and who are "fighters." We reward forceful management and quick problem resolution. We favor intellectual simplification and clarity. Although we know the dangers of overconfidence, we emphasize building internal scripts that reinforce confidence rather than assessing carefully the limits of our knowledge. Such conviction habits reflect the assumption that we typically know enough based on the past to decide how to proceed. But as we move toward release in complex interconnected global systems, inductive reasoning based on experience during growth and conservation phases of the adaptive cycle becomes less reliable. In this chapter, we examine the underdeveloped skillful habits of humility, and we explore how humility and softened forms of conviction can be integrated so we can flourish now.

Consider a puzzle that reveals internal tensions in our views about conviction and humility. Politicians are often criticized for being "flip-floppers"—that is, for changing their minds. For example, when John Kerry was a presidential candidate in 2004, he was effectively ridiculed for being a flip-flopper because he appeared to have changed his mind about supporting the war in Iraq. President George H. W. Bush was similarly criticized when he changed his position against raising taxes. To many, these examples suggested lack of conviction. At the same time, scientists are praised for changing their minds when new evidence supports an alternative view. Albert Einstein believed in an eternal universe and rejected the early theories regarding a "big bang." Yet when Edwin Hubble established that galaxies were rapidly moving away from each other, Einstein changed his view well before the Big Bang became commonly accepted. Did he flip-flop? We do not criticize scientists for changing their minds based on new evidence; indeed, this change reveals the virtue of humility. Why should our leaders be praised for their convictions and scientists for their humility? And can these two positions be reconciled, or are our social norms regarding conviction and humility hopelessly inconsistent?

One possible explanation of this puzzle is that we want to see conviction about moral views, and we praise humility about empirical matters. On such a view, our leaders' actions should be based on their moral commitments, and scientists should be responsive to changes in empirical evidence. But this simple explanation is not adequate. We do admire steadfastness regarding some moral matters, and especially about moral character traits, but such steadfastness rarely dictates political actions. Whether we should support a war, raise taxes, or promote renewable energy depends a great deal on the specific empirical evidence for these policies. Politicians may base their general views on broad moral principles, but as they apply these in specific context by advocating for policies, they must depend on the kind of evidence that scientists use. Thus, as the evidence politicians have changes, it seems reasonable that their views should change, just as they do for scientists.

A more promising alternative explanation is that effective leadership requires conviction. We want politicians to be effective leaders who can marshal support from others. Simple, clear, consistent messages tend to garner support, whereas complex, situation-dependent answers tend to confuse followers. In a democracy, leaders are elected based in part on the policies they advocate, so if they change these policies, they may be guilty of misleading their supporters. Such changes may also mislead their allies and embolden adversaries. Moreover, we are particularly concerned that politicians are subject to changing their views for bad reasons, not because the evidence has changed. For example, they bend to shifts in popular opinion or respond to forceful lobbying from powerful special interests. Such shifts often seem to betray weakness or lack of integrity. Convictions, specifically beliefs that others can count on, are particularly important for leadership roles.

Most of the time, scientists are not trying to lead a populace; they are primarily interested in what is true. The evidence for what is true may be unclear, complex, or changing, so scientists typically respond in kind. The role of scientist is subject to much less social demand for consistency over time or certainty at a time. The nature of science requires responsiveness to shifting evidence and acknowledgment of uncertainty. We want scientists to be open to new hypotheses that might better explain the evidence.[1] Many of our greatest scientists, like Einstein, are exemplars of humility.

On this account, differences in the legitimate social demands associated with the roles of political leader and scientist create the puzzle. Although this may be a partial explanation, it cannot be the whole story, because it is still unclear how a leader can appropriately have policy convictions when the policies must be in part evaluated on the basis of changing scientific evidence. It now appears that our norms for leadership are in tension with wise leadership decision-making (which requires more humility). We should want leaders who are willing to change their minds when the evidence warrants it. A more complete explanation must await our discussion of how we can integrate conviction and humility.

This puzzle reveals one tension between conviction and humility, but there are others. Shared convictions can reinforce social bonds and simplify group actions. Convictions also reduce uncertainty and bolster self-esteem. Having convictions tends to feel good; sometimes we reasonably want the stability they provide. Yet convictions can also polarize a society, dividing it into competing tribes. By contrast, humility can build bridges across cultural differences, open us to alternative ways of living, fuel our curiosity, and speed our growth. We must balance the benefits of conviction and humility if we are to flourish.

These tensions do not mean humility and conviction are opposites. Indeed, humility is better juxtaposed with arrogance. Convictions can be arrogant, but they need not be. Nonetheless, developing the skillful habits associated with conviction is often at odds with the practice of developing humility skills. Unfortunately, our cultural institutions give undue weight to conviction in our lives. This results in too many beliefs becoming firm convictions that we are unwilling to question, and too much emphasis on conviction skills. We need to build skillful habits of humility strong enough to balance and temper the cultural emphasis on conviction.

Our Conviction Conveyor Belt

Our current political polarization is a salient manifestation of a conviction-oriented culture. On both the right and the left, viewpoints have hardened, and compromise has become increasingly unattractive.[2] Key votes in Congress are divided largely along party lines. Even when bipartisan legislation passes, it is often underpublicized. Powerful feedback loops between political leaders, media, and the politically oriented populace serve as conviction generation machines that transform ordinary beliefs into identity-forming "undeniable truths."

An insightful book by Michael Lynch, *Know-It-All Society*, examines the personal and cultural dynamics that result in the multiplication of convictions, creating an age of "intellectual arrogance."[3] Lynch characterizes a conviction as "a commitment that reflects the kind of person we want to be."[4] We can envision a continuum between a mere belief that we could easily change, through an emotional commitment that begins to become part of our identities, and on to self-defining commitments that we cannot imagine changing. Key institutions in our society, especially those connected to morality and politics, tend to create conveyor belts that move ordinary beliefs up the continuum to hard convictions. This has happened to views about climate change, which increasingly form litmus tests for whether one belongs to the left-wing political tribe or the right-wing tribe. Many on the left are convinced that we are in a climate crisis, while many on the right are convinced that climate fears are overblown at best and at worst part of a deliberate hoax designed to expand government interference in the market. The irony is that beliefs about the existence of climate change seem to be paradigm cases of empirical beliefs best settled by scientific evidence. But their entanglement with tribal identity has moved them up the conviction conveyor belt.

Robert Abelson complicates Lynch's picture of convictions.[5] He identifies three dimensions along which convictions may vary: emotional commitment, ego preoccupation, and cognitive elaboration. The emotional commitment dimension is most evident in the discussion of flip-flopping and in Lynch's analysis. It includes the degree to which a belief is part of our identities, its degree of certainty, and its likelihood of changing. The ego preoccupation dimension includes the strength of the belief, the frequency of thinking about it, and its importance. These two dimensions overlap, but we can take a belief to be certain and unchanging without believing it to be very important or thinking about it often. For example, I am quite sure that there are dust bunnies under my bed, but I rarely think about it and don't care much. The third dimension, cognitive elaboration, pertains to how well we have worked out the implications of a conviction, and how long we have held it. For a belief to be a conviction, it must be relatively high on at least two dimensions.

A conviction that climate change is the most pressing concern facing humanity nicely illustrates the way Abelson's three dimensions can affect our lives. This conviction often involves high ego preoccupation. It is highly salient and strongly held. It often leads to behavior choices like using mass transportation and renewable energy. For some people with this conviction, though, cognitive elaboration may be low; their understanding of the details of climate science may be hazy. As a result, they avoid contentious conversations about climate change and prefer to talk with peers who share their conviction. Although high emotional commitment may be associated with high ego preoccupation, it need not be. A conviction about a climate crisis may not be part of their identity and may be quite changeable. Next month, human population growth, or inequality, or biodiversity loss may seem much more important, and it may motivate different behavioral changes. Our conviction conveyor belt sometimes leaves old convictions in the dust as other issues receive more attention. In just a few years, immigration issues have become highly salient, generating widespread competing convictions in a fairly short time.

Our media system plays key roles in our cultural emphasis on conviction. As many have noted, the fragmentation of our media along ideological lines means that most of us are exposed to stories that support our worldviews. We tend to select our news sources, such as Fox News, MSNBC, or public radio, in part because their stories reinforce our beliefs.[6] Increasingly, though, the American public accesses news media online. Since most online media make their money through advertisements, they need to maximize page views, which can be achieved with more sensational depictions of disasters resulting from policies supported by our opponents or by directly demonizing these opponents as ignorant, out of touch, or corrupt. Popular commentators are often role models for having strong convictions. Humility is rarely in evidence; it simply does not make good copy. We still find lots of good, fact-based reporting online, and typically some balance in viewpoints that are expressed in major media outlets. I do not want to overstate the power of the media's contribution to the conviction conveyor belt. But given the structure and economics of our major media sources, the tilt is toward belief reinforcement.

Social media is in many ways worse. As Lynch points out, platforms like Facebook and Twitter (restyled as "X") are designed to build emotional engagement, not intellectual growth. They achieve this by enabling bonds with like-minded people and populating feeds with media that will appeal to us given our prior preferences. They can do very well at fostering bonding social capital, in-groups that share key values and viewpoints. But they struggle to support the expansion of bridging social capital, interactions across bonded groups. The more we are exposed to views that are like our own, the more our viewpoints are reinforced and become strengthened. This is not irrational. Much of what we believe is based on the testimony of others, such as scientists, teachers, doctors, and peers. Agreement among those whose testimony we respect tends to increase our confidence in that testimony. Consensus on a belief certainly does not show it is correct; logicians would call this the bandwagon fallacy. It does lend credibility to testimony, however. It is one of the ways we test testimony. Studies on group polarization also indicate that after discussion in homogeneous groups, viewpoints tend to become more extreme.[7] For example, when most jurors generally agree regarding an outcome in civil cases, they tend to give higher awards than the individuals would have given in advance of discussions. Not only do our beliefs become stronger; they tend to become more extreme as a result of the kind of media environments that surround us.

Insofar as the institutions of business and politics idealize forceful leaders who model strong, emotionally committed beliefs, they too contribute to our bias in favor of conviction. Indeed, the feedback loops between education, media, and leadership models reinforce a picture of how we should approach problems—with the clarity, conviction, and confidence that will effectively move people to act. Although confidence and conviction are separable, they are close cousins. For example, confidence about our belief-generating mechanisms is involved in conviction, and high self-confidence tends to be associated with a strong sense of identity, including identity-forming beliefs. Innumerable books about success talk about how to build self-confidence. Confidence is seductive, powerful, and intoxicating. Our cultural love affair with confidence, with the "can-do attitude," is yet one more contributor to the conviction orientation that competes with cultivation of humility skills.

The role that media and politics play in ratcheting up conviction became highly salient as the COVID-19 pandemic spread. Information about the dangers associated with the spread of the virus was fragmentary and evolved very quickly. In the early phase, scientists and public health officials emphasized that we did not have enough evidence to form firm beliefs about how contagious the virus was or how likely it was to kill people in different groups. But understandably, people felt discomfort with the uncertainty that was warranted by our situation; politicians and media personalities responded with more authoritative views, which often reflected their political beliefs. On the right, people tended to downplay the dangers, and on the left, there were calls for swift government action. Politicized narratives regarding the virus threat began to form, and these influenced information flow throughout the pandemic.

As infections and deaths rose, the majority of the country entered lockdown. Still, confirmation biases tended to reinforce ordinary people's answers to questions about whether this lockdown was necessary. Those who were focused on the economy or on individual freedoms were skeptical about a lockdown and urged lifting it quickly. Others concerned about vulnerable populations thought that lockdowns came too late and that government actions were insufficient. Throughout the pandemic, only anecdotal evidence could be acquired firsthand; all other information needed to be filtered through testimony, usually from sources selected because of fit with preferred ideological orientations. The rational thing for most people to do would have been to double down on humility—to recognize how much is unknown and to avoid firm belief, to be patient about lack of clear guidance from officials, and to make decisions based on their general approach to decision-making under uncertainty. But because we live in an age of media fragmentation and increasing distrust in social institutions, divergent, quickly formed beliefs became hard convictions, and it was difficult to marshal support for a unified approach to the pandemic.

Even our education system, which could be a humility magnifier, seems conviction-oriented. Take, for example, a standard approach to teaching and assessing critical thinking: the thesis defense paper or speech. When completed efficiently, this kind of assignment usually leads to cherry-picking evidence to support an antecedent viewpoint. Even when such an assignment requires consideration of alternative views and objections, we have to be honest and admit that we generally receive just a more sophisticated selection of confirmatory information. Such a paper rarely results from a genuinely balanced investigation, even if it does teach some argumentative skills. We want students to learn argumentative skills, but it would be even better if they also learned how to provide a careful assessment of the strengths and weakness of the view they are inclined to accept. If we focused more on developing humility in the classroom, the process of educating for critical thinking would focus more on how to navigate uncertainty while gearing the degree of belief to the evidence. Though some of this does happen, it is far from the norm, in my experience. Our standard educational practices, including those associated with the movement to build youth self-esteem, create a bias in favor of conviction. This tendency is amplified where ideological rigidity among some students and faculty reinforces common beliefs.

Social forces are not the only causes of our conviction emphasis. We are all predisposed toward reinforcing our beliefs. A sizable literature on confirmation bias shows how we tend to seek out evidence that supports our views and ignore inconvenient evidence that undermines them.[8] For example, in the famous Wason experiment, participants were given a short series of numbers and asked what the rule was that generated the series.[9] They formed hypotheses and then tested these by asking whether another series fit the rule. A significant majority used only series that would confirm their hypothesis, which typically prevented them from finding out that their hypothesis was false. Confirmation bias is one of many nonrational belief reinforcers. We also tend to resist change, so that once we have invested in a belief, an inertia sets in that makes us reluctant to abandon it.

My point in this section is not just that our institutions tend to increase our *motivations* for forming convictions in ourselves and others; they cultivate *skills* that build conviction and confidence. Such skills include the techniques often featured in confidence-building books for shifting our self-image—our internal monologues and our self-presentation. They include methods like those found in marketing texts for building emotional connections, and techniques taught in most composition classes for developing a sustained argument. More advanced skills deal with how one creates communities that reinforce and strengthen emotional commitment and move belief into tribal identities. President Donald Trump very skillfully used in-group/out-group dynamics to reinforce convictions and undermine the effectiveness of criticism by the use of ridicule, ad hominem arguments, and colorful distractions. We can appreciate these genuine skills whatever we believe regarding the use to which they are put.

To be clear, I am not saying that everyone has too many convictions or too much self-confidence. Convictions are important. Some people with too few convictions seem unmoored and are often unmoved by crises we face. Our problem is closer to Yeats's famous couplet in "The Second Coming": "The best lack all conviction, while the worst / Are full of passionate intensity."[10] My claim is that key cultural institutions overemphasize conviction in our lives, and that results in too many beliefs becoming hard convictions that we are unwilling to question or change. When that happens, we have too much "passionate intensity," and it becomes hard to change direction when the circumstances require it and to work effectively with those who have different convictions.

Humility and Flourishing Now

Like conviction, humility has multiple dimensions of meaning. The recent literature features more than a dozen different accounts of humility and its close cousin modesty, but these can be roughly divided into two groups. One group identifies humility with our self-assessments, the second group with our approach to social interactions. Self-assessment views include owning our cognitive and moral limitations, acknowledging our strengths (without overemphasizing them), and appropriately assessing the evidence for our beliefs. Social interaction views focus on how we treat others, especially those who are differently situated or who have different beliefs. They include having low concern for status and entitlements, focusing primarily on others rather than ourselves, and being open to others' viewpoints. I see these accounts as highlighting different clusters of humility skills rather than forming competing accounts of what humility is.

I divide humility skillful habits into three broad categories: those skills that directly soften conviction, those involved in the reflective practices that assess our limitations, and those that help us to focus on others rather than ourselves. The conviction-softening skills prevent beliefs from moving up the conviction conveyor belt too far. They enable us to avoid undue emotional commitment and ego preoccupation with beliefs and thereby make us more cognitively flexible. Reflection skills strengthen our abilities to accurately assess our strengths and weaknesses, which has an indirect effect on our levels of conviction. Decentering-self skills enable us to be more open to and focused on others' perspectives. Box 2 summarizes the main humility skills I discuss in this chapter. Before exploring these skills, we need to see why our current cultural emphasis on conviction is so problematic—why we need to cultivate stronger humility skills.

Box 2. Humility skills

Soft-conviction skills

- Exercising skepticism regarding conviction hardeners
- Moderating emotional commitment and other elements of conviction
- Communicating nuances in strength of beliefs

Reflection skills

- Identifying and owning our cognitive limitations
- Appreciating our strengths
- Effectively using our mistakes
- Developing sustainable reflective practices and peer groups

Decentering-self skills

- Prioritizing attending to others
- Exercising genuine curiosity about alternative viewpoints
- Reducing concern about status and entitlement
- Feeling gratitude for contributions of others

One danger of an overemphasis on conviction is that we become intellectually arrogant, which Lynch defines as being certain that our viewpoints are correct and that we have nothing to learn from other viewpoints.[11] Conviction alone need not have this implication, but when conviction becomes part of the identity of a tribe, and the tribe sees itself as superior in light of its correct convictions, the social reinforcement cycles lead to intellectual arrogance.[12] When we believe listening to our opponents is pointless and compromise untenable, then we view issues as zero-sum games. As we saw in chapter 2, where such a competitive approach to problem-solving dominates, we have little capacity to forge creative win/win solutions that are necessary to address many of our challenges. If tribes are roughly evenly matched, gridlock becomes common, and in worst-case scenarios, violence seems the only viable exit door. Humility enables us to see the limitations of our tribe and weakens identity based on conviction.

Intellectual arrogance is not just a barrier to effective collaboration; it also leads to misguided assessments of our limitations and thus more cognitive mistakes. Convictions about the behavior of complex systems that we are destabilizing are rarely warranted. We have seen that thresholds beyond which a system changes rapidly are often hard to identify in advance. We often fail to understand important feedback loops. Furthermore, small-scale perturbations in a system may unexpectedly have large impacts on system dynamics (through the "butterfly" effect). For example, we may model how ice cap melting may affect ocean currents, but we typically do not know which model best represents thresholds that govern shifts in the Gulf Stream. Under these conditions, having convictions about how likely we are to shut down the Gulf Stream seems unwise.

Of course, we must have beliefs about these systems in order to act, but we should be acutely aware of the limitations of our understanding and the dangers of using ideology to fill gaps in our knowledge. We may have ample justification for beliefs about causal interactions of some elements of a system—for example, that increasing the use of renewable energy will decrease total carbon emissions. But that does not mean we have nothing to learn from people who disagree with us on the wisdom of renewable energy incentives. People differently situated within the nested systems affected by incentives may see potential unintended side effects that should be taken into consideration. As we saw in our initial puzzle, once we move from general principles to their application in a context, hardened convictions are much harder to defend. As we grapple with the systemic dimensions of more wicked problems, our flourishing will be enhanced by greater open-mindedness and fewer hard convictions. Humility enables us to fine-tune our epistemic evaluation of beliefs.

Further, insofar as the United States is moving from the growth/conservation phases of the adaptive cycle through release and reorganization, humility protects us from being too bound by past experience. Convictions about systems behaviors and how to manipulate systems effectively are acquired through induction from past experiences or testimony from others whose inductions we trust. Thus, holding strong convictions about how to respond to a system dynamic requires the assumption that we know enough based on the past to decide how to proceed. But when systems structures are changing dramatically, induction regarding system behavior is less reliable. Prudence demands more humility and less conviction. We should avoid the hubris involved in generalizing the knowledge that we do have beyond its legitimate scope.

Intellectual arrogance also suppresses creativity. The release and reorganization phases of the adaptive cycle put a premium on creativity. We must consider more "out-of-the-box" solutions to our challenges in order to develop novel prototypes that may successfully lead to fruitful ways of reorganizing the system. Creativity is important for navigating all phases of the adaptive cycle, but the breadth of creativity must expand in the back-loop phases. The more convictions we have, the more our creativity is constrained. If we just seek solutions compatible with unearned convictions, we will miss promising novel solutions. The history of science contains many such examples. Among the most famous is how the conviction that planets must have circular orbits made it very hard to reconcile the new Copernican heliocentric model of the solar system with increasingly precise data regarding planetary movement. Only when Kepler abandoned the conviction and explored elliptical orbits could the heliocentric system make significant progress. Humility bolsters creativity.

Finally, our tendency to follow those who demonstrate conviction creates special problems for leaders; they face significant pressure to claim that they know how to proceed in circumstances that generally make such knowledge claims unwarranted. For example, in 2014–15, President Obama confronted a complex situation in the Middle East. The Islamic State (ISIS) expanded its activities in Syria and Iraq and created affiliates in other countries that orchestrated massive terrorist attacks, notably the attacks in Paris where over 130 people were killed. He was severely criticized in late 2015 when he said that he did not yet have a strategy for defeating the Islamic State and that he was consulting military advisers, allies, and Congress before determining his strategy. His statements were honest and probably reasonable given the uncertainties about what strategies would be successful, but they also violated the strong social norm that good leaders should know what to do. To many, he looked weak and unconvincing as a leader. In such situations our norms encourage self-deception or hypocrisy in our leaders. Leaders must either ignore the uncertainties and come to believe that they do know what to do, or they must consciously say they know when they do not. Neither option is attractive. Over time, both hypocrisy and self-deception result in lack of trust in the institutions that we need to help us address our challenges.

Flourishing in the age of climate change will require more cognitive flexibility, more creativity, and more careful assessment of the justification we have for our beliefs. Where increasing conflict tends to harden conviction, we will be better off if we exercise the skills of softening conviction. When rapidly changing circumstances upend familiar ways of addressing problems, we will be more likely to flourish if we become more comfortable adopting experimental approaches to decision-making—testing novel ideas rather than working through the implications of convictions. We must be ready to change our minds quickly and to listen to diverse perspectives in order to find our blind spots. Strengthening humility skills will help each of us to flourish, and if enough of us strengthen these habits we will be more likely to reach the collaborative solutions to problems that approaching sustainability requires. Of course, humility is important for flourishing during any timeframe; there are general reasons why humility and convictions skills should be in balance that I have not surveyed. My point is that we are seriously out of balance and that managing the challenges and opportunities associated with the next thirty years will require an emphasis on humility skills.

Integrating Conviction and Humility

Since our current emphasis on conviction makes humility seem like weakness, it is hard for us to see how conviction and humility can be integrated. We need to see how both sets of skills can be used together in the appropriate situations and how both can be emphasized in our decision-making. We need a version of binocular vision here that integrates the skills associated with each. To see how this is possible, we need to draw a rough distinction between hard and soft convictions. "Hard" convictions are unyielding. They are treated as certain and important enough that they cannot be compromised. They are not up for debate, and they do not shift when our evidence changes. In other words, they involve high ego preoccupation and emotional commitment.

By contrast, "soft" convictions serve as a stable guide for behavior, and they involve enough emotional commitment to motivate group action, but they are usually far from certain. We may identify with a soft conviction, but that portion of our identity is modifiable under the right circumstances. We can compromise regarding our soft convictions. Soft convictions tend to have a moderate degree of ego preoccupation and emotional commitment. They serve to fulfill the roles played by conviction in our identities and in leadership, but we understand them to be modifiable if we acquire information that makes them unworkable.[13] Soft convictions are more easily integrated with humility.

Soft convictions occupy a space on a continuum between mere beliefs and hard convictions. Any specific conviction may be harder with respect to one of Abelson's three dimensions of commitment and softer with respect to others. Both hard and soft convictions may range widely with respect to degrees of cognitive elaboration. In general, the more we understand the implications of our views, the more reasonable it is to strongly hold a view if our investigations leave it well justified. However, frequently the more we know about a topic, the more we run into uncertainties, which warrant softer convictions. Unfortunately, in our culture, many hard convictions involve little cognitive elaboration. Ironically, the less one knows, often the easier it is to hold firmly to a core belief.

The conviction conveyor belt tends to harden conviction prematurely and render us less likely to assess our beliefs realistically. It rachets up emotional commitment and certainty in beliefs that have not earned that status—that are not sufficiently justified. Some beliefs are reasonably treated as hard convictions, but most should remain mere belief or soft convictions. Core moral beliefs or highly justified empirical beliefs may legitimately become harder convictions. For me, and many others, the belief that all people should be treated with equal respect is an appropriate hard conviction. An expert in a field who has invested tremendous effort into elaborating a viewpoint may reasonably form hard convictions regarding its central tenets. The problem is not the presence of hard convictions, but rather their overpopulation.

To combat the conviction conveyor belt, we need to strengthen conviction-softening skills. The first of these involves becoming skeptical about social conviction hardeners. Once we appreciate the social mechanisms that harden convictions, we can learn to note their presence and to be suspicious about their impacts. If I hear a commentator ranting about some egregious activity, my default reaction should be to raise questions about the claims. If I am surrounded by compatriots who are emotionally affirming some view, that situation should trigger me to think critically about its validity, rather than to join the bandwagon, even when it fits my biases. I am not suggesting that we should doubt everything we hear; most of what we know relies on the testimony of others. We should become good at thinking about alternative explanations for events that have strong emotional valence and to be suspicious about how the strength of our convictions might be manipulated or socially influenced.

Conviction softening also involves deliberately moderating our degree of certainty, our levels of emotional commitment, and the amount of attention accorded to a belief. Although we may not regularly deliberate about these levels, they are at least partially within our control. With some practice, we can choose how we allocate our attention. Attention can then be used to soften convictions. For example, when contemplating models used to predict the effects of climate change, we can choose to focus on potential areas of uncertainty regarding climate models, or we can focus on their shared implications. We can focus on worst-case scenarios, or we can focus on models nearer to best-case scenarios. Attending to each of these is important, but if I know that I have a tendency to focus on worst-case scenarios, I can choose to focus more on other scenarios.

Similarly, we have a degree of control over where we allocate emotional energy. When the latest outrage-inducing story about a politician's actions tends to strengthen emotional commitment, I can learn to pause and bracket my emotions so I can assess their appropriateness. I may decide to believe the story without the fervor that would make it hard for me later to modify my beliefs if new evidence arises. These subskills of managing attention and commitment may also be used to bolster conviction, where undue focus on uncertainty makes us unwilling to commit to action. The key is to tune the skills so that soft convictions become reasonable stopping points for deliberation.

Another group of conviction-softening skills involves our abilities to communicate nuances regarding our beliefs. Being able to signal our degrees of commitment, certainty, preoccupation, and cognitive elaboration with respect to a conviction enables us to build trust within a group and to increase its capacity to act with flexibility. For example, leaders must often communicate enough commitment to a strategy to motivate followers to act without implying degrees of certainty that later lead to charges of hypocrisy. During World War II, Churchill communicated a powerful commitment to the cause of defeating the Nazis, but early in the war he avoided implying certainty that it would be accomplished. A similar position with respect to climate change seems reasonable. Sometimes, we may need to communicate a high degree of certainty about a principle but a much lower degree of certainty about its application in a context, without signaling a lack of commitment to the principle. In any contentious decision-making process, a leader's communication has at least two audiences: those with whom the leader disagrees and those the leader represents. The skills of driving hard bargains while leaving open flexibility usually involve subtly communicating degrees of conviction.

Some might object that expressing uncertainty or moderated emotional commitment about beliefs weakens our case for action. Indeed, the opponents of climate action have tried to use uncertainty to undermine the case for action, arguing that we should not make the economic sacrifices involved in mitigating climate change until we know for sure that they are necessary. This example reinforces the view that we need to express hard convictions about the facts to convince others to act now. However, soft convictions can also have a powerful impact on action.

Rachel Carson's approach to highlighting scientific uncertainty in *Silent Spring* provides a powerful example of using key soft-conviction skills to influence people. An analysis of her drafts of *Silent Spring* conducted by Walker and Walsh indicates that she quite deliberately used scientific uncertainty to amplify the risks of pesticides like DDT and to invite the public to participate in the debate about their growing use.[14]

By the time she embarked on the research that would become *Silent Spring*, Carson was a well-respected biologist and science writer. She had served for fifteen years in the US Bureau of Fisheries, which became the US Fish and Wildlife Service (FWS), eventually becoming editor in chief for all FWS publications. She had written three popular books on the natural history of oceans. In her time as a public servant there is no evidence that she relished advocacy or sought controversy. Yet her deep concern regarding the potential impact of widespread pesticide use led her to write a book late in her life that would jump-start the environmental movement and make her one of its world-renowned leaders.

In post–World War II America, the public had enthusiastically embraced the role of science in developing new technologies that would improve quality of life. Carson was acutely aware that *Silent Spring* would buck that trend. She also knew it would generate a significant backlash from the chemical industries that made pesticides. Thus, she had to be careful not to overstate her case and provide detractors with ammunition for undermining her arguments. No doubt this was part of her motivation for being cautious in how she framed her evidence, but this does not explain either why she played up uncertainty increasingly in later drafts of the book or why that uncertainty did not have the effect of undermining her case for action. Walker and Walsh find thirty-three distinct places where Carson highlighted uncertainty using phrases such as the following:

> No one knows what the ultimate consequences may be.

> The whole concept of genetic damage by something in the environment is also relatively new, and is little understood except by the geneticists, whose advice is too seldom sought.

> For this reason the role of chemicals in general use (rather than in laboratory experiments) had not yet been assessed. It is extremely important that this be done.

> Scientists do not agree upon how much DDT can be stored in the human body.[15]

Throughout the book, Carson demonstrated the skills of carefully framing the degree of scientific evidence we have regarding the safety and potential negative impacts of pesticides. At the same time, she vividly described cases where negative impacts are evident, and the cumulative effect of such cases is telling.

Moreover, she moderated her expressions of belief and emotional commitment when confronted with contrary evidence. She had many experts read drafts of her chapters; Walker and Walsh show how she reframed key points after criticism from scientific experts in the field. For example, she sought Dr. George Criles's views on her chapter about the relation of environmental toxins to cancer, and Criles was quite critical of her recitation of the evidence, calling it "overwritten" and suggesting that she was speculating.[16] As a result, she added a paragraph in the published version indicating that "medical opinion is divided on the question" but continued on with evidence supporting her view.

Another indication of soft-conviction skills was her framing of the actions to be taken as a result of her evidence. She does not call for an outright ban on DDT and other pesticides; that would come later. Her conclusion is tempered and flexible. She advocated that such pesticides be used only when other methods would not work. Instead of presuming that after all of her research she has the expertise to decide what should be done, she invites the public to weigh in based on its values after considering all the evidence. As Walker and Walsh put the point:

> Uncertainty's dual function of disruption and actualization gives lay readers the rhetorical grounds on which to draw their own inferences and assert their right to know the potential hazards of environmental situations. It also loosens the ethical handcuffs on the scientific writer.

The writer's ethos changes from an authoritarian voice ("Let me tell you what is") to a more collaborative one ("Let me tell you what we don't know and what might happen"). Uncertainty helps fulfill the prophecy of Carson's book that the public will decide for themselves.[17]

The public would have no role in a debate between scientists from the chemical industry and environmental groups, but by highlighting scientific ignorance, Carson creates room for the public to participate in policy formation by showing how values must carry the weight of argument in cases of uncertainty. Clearly Carson has significant emotional commitment to her views that we needed to change our practices of pesticide use, but she felt just as strongly that a collaborative public process was required to determine what policy is appropriate.

Carson closed *Silent Spring* with a critique of the arrogance associated with the control-of-nature philosophy that underlies widespread pesticide use to control insect outbreaks. The whole book is a call for humility before the complexity of nature and the uncertainties of our best science. Here we see leadership with passion and commitment but without the hard convictions that often undermine collaboration, creativity, and responsiveness to changing evidence.

The impact of *Silent Spring* was immediate. It was serialized in the *New Yorker* in the summer of 1962. The full book was published in September of that year, and it was selected as the Book of the Month for October, so it reached a large audience quickly. The anticipated harsh reaction from the chemical industry only served to increase its readership. Public opinion and that of much of the scientific community landed squarely on the side of restricting pesticide use within a year of its publication. By the early 1970s DDT was being phased out by the newly created Environmental Protection Agency.

Carson's example illustrates how the use of soft-conviction skills can both stimulate action and help to avoid mistaken hard convictions that can backfire and reduce both trust and persuasiveness. She makes it clear that tentativeness is not to be equated with timidity; she is quite forceful in her advocacy. Her example suggests that she has developed a robust binocular vision that enables her to integrate moderate forms of conviction with a humble approach to her subject. This is a balancing act in which the possibility of changing her mind in light of new evidence and the emotional commitment to her concerns are both at the forefront of her thinking.

Reflection and Decentering the Self

The two other clusters of humility skills I highlight contribute to softening convictions, but I will focus primarily on other ways they enhance flourishing in the age of climate change. The skills of reflection are central to self-assessment views of humility mentioned above, and the decentering-self skills are associated with social interaction views of humility. The three humility skills clusters are mutually reinforcing, which partially explains why they are taken to be expressions of a single virtue.

The reflective skills associated with humility all involve stepping back from what we are doing and viewing it from a distance, thinking about our actions and their context from alternative perspectives, and frankly assessing ourselves. Engaging in these practices often enough and with increasing accuracy helps us avoid mistakes and learn from those we make. When the context for our action is novel or complex, reflection becomes even more important. This is especially true when the costs of error are high and the changes we must consider are great. We have seen that the age of climate change is a context where all these conditions apply. As we near release and reorganization on large scales, we must reflect more often and more astutely.

Perhaps the most important reflective skill involved in humility is recognizing our limitations or weaknesses and keeping them in mind as we act.[18] Humble people are acutely aware of the weaknesses of their belief-forming processes. They act on that recognition by limiting their emotional commitments to beliefs. They recognize biases and work to rectify them. Such reflection reinforces soft-conviction skills. In a period of accelerating change, the limitations of our understanding become more pronounced. When our local systems are strongly impacted by the dynamics of larger systems nearing release, the uncertainties governing any action are magnified. To navigate the resulting turbulence, we must be acutely aware of the limitations in our understanding and yet maintain our ability to act and to change course when that is warranted.

A one-sided focus on our limitations, however, would likely lead to indecisiveness and low self-confidence. We must also reflect on our strengths—our genuine knowledge and our capacities to act. Such reflection can help us take appropriate pride in what we do well. Pride in this sense is distinct from intellectual arrogance. It does not involve certainty about our viewpoints or unwillingness to learn from others. Historically pride is sometimes understood as the sin resulting from lack of humility. So understood, pride inevitably involves failure to appreciate our limitations. Most current writers about humility acknowledge that reasonable pride is important and that it helps one to avoid the risks of too much humility.[19] The skills of reflecting on our strengths serve as an antidote to an overemphasis on our limitations. When uncertainty is high, we must rely more heavily on our sense of character, but that character must be grounded on realistic assessments. As we saw, Rachel Carson was acutely aware of the limitations of her knowledge about the long-term impacts of pesticides, though she was rightfully aware that she still had a powerful case to make about their dangers. She knew the strength of these arguments, and the knowledge buoyed her as she took on the critiques of the pesticide industries.

Another reflective skill involves learning from our mistakes. No one likes to make mistakes, especially where large mistakes can have dramatic consequences. In this turbulent time, when inductions from the recent past are less reliable than in more stable periods, more mistakes are inevitable. To flourish, we must become more comfortable making small mistakes and rapidly learning from the results. Strengthening our capacity to take an experimental approach to decision-making—trying a possible solution on a small scale, evaluating its impacts, and adjusting the intervention to make it better—is one way to systematize learning from our mistakes. It does not apply in all circumstances, but it is a skill that is especially important when uncertainties are very high, as in the release and reorganization phases of the adaptive cycle. This practice is one of the best ways to assess how well we are balancing our focus on our limitations and our strengths.

Often, however, we cannot avoid our blind spots without the help of others who are willing to take the time to highlight our deficiencies. We need to get better at welcoming such criticism into our lives. We all tend to join groups of people who think alike. We feel at home in groups that reinforce our strong beliefs. As we have seen, this tendency, amplified by media that support our perspectives, fuels the conviction conveyor belt. It is less comfortable, yet more fruitful, to have some friends that challenge our beliefs and blind spots, but do so in a way we can hear. We need to cultivate friends who challenge us productively and work on becoming such a friend to others with whom we are close. We will return to this theme at the end of the chapter.

To strengthen our community of productive critics we need to foster the third large cluster of humility skills, which involve different ways of decentering the self in one's interactions with others. Humble people are more interested in hearing from others than from themselves; they avoid centering their attention on the self. They have cultivated the skills of focusing primarily on others when they are interacting—listening well, attending to body language, being sensitive to ways that context shapes behavior. These skills reinforce the empathy skills we discussed in the last chapter. They are particularly important during periods of release when we need to build new relationships quickly and when conflict is heightened. Humble people are very good at being fully present when they interact with someone. They are curious about people. Arrogant people quickly lose interest when someone else is talking. Their eyes dart around, or they begin thinking about what they want to say.

In general, our ability to exercise curiosity assists in decentering ourselves and focusing on other people or parts of our environment. While we tend to think of curiosity as a trait that is not under our control, Todd Kashdan and others have done research that shows how we can strengthen curiosity and why it might be wise to do so.[20] We can learn how to ask good questions that elicit engaging answers. We can reframe a potentially boring situation so that details about group dynamics or behavior motivations become more interesting. We can learn how to draw wisdom from those who have experiences very different from our own. In short, we can develop the skills of learning from a wide range of human interactions.

Curiosity is crucial in times of rapid change, when having a learning mindset enables us to adapt quickly as circumstances shift and to experiment with innovative solutions to challenges. Instead of focusing on what we know, we focus on what we do not know but could learn through our interactions with the world. Kashdan has developed a multi-factor model of curiosity that identifies five habitual patterns of behavior, each of which can be modified to build curiosity skills. These include the social curiosity described above, and also our ability to tolerate the stress of novelty, our capacity to experience joy in exploring new aspects of the world, our tendencies to seek thrills, and our sensitivity to information deprivation. He has used this model to look at ways that organizations can cultivate cultures of curiosity.

Another skill of decentering the self involves ignoring any status and entitlements that our achievements may warrant. Arrogant people usually want others to acknowledge and reinforce their status. It may seem natural that people with PhDs and numerous publications have higher status than undergraduates in an academic setting. They may thus expect to speak first and at length about topics within their expertise. But humble professors interact with a group as if such status was irrelevant. Even if entitlements are proffered, they prefer to hear the views of others, offering their own views late in the conversation and without fanfare. We must cultivate the skills of selectively ignoring our own status in order to welcome fully all members of a group and to encourage alternative perspectives and productive critiques. When old status hierarchies are eroding, we are more likely to flourish if we are not focused on preserving our status.

The literature on humble leadership in business often features CEOs who so successfully decenter their selves that they become highly accessible to anyone in the company with a good idea or a concern.[21] Mary Powell, the CEO of Green Mountain Power between 2008 and 2020, used the symbol of physical space allocation to signal this kind of humility. She eliminated the ornate upper-floor executive suite protected by two levels of executive assistants and instead occupied a desk in a cubical within an open workspace on the first floor in the customer service area. There she could interact regularly both with customer service agents and with customers who had issues.[22] The symbol was clear: we are here to serve the interests of our customers. The leaders do not have all the ideas for how to do this better; those ideas often come from those on the front lines of the business. This kind of humble servant leadership model has a long history, but it still seems to be relatively rare in a society that exaggerates the importance of status, celebrity, and conviction.[23]

The last decentering-self skill I will mention is acting with gratitude. When we approach other people with gratitude, we focus primarily on what they bring to us. We avoid taking them for granted. Gratitude opens us to the world and has a positive impact on our flourishing, our mental health, and our physical health.[24] If we are humble, we typically think that the positive things that happen to us are not fully deserved—they are gifts. Gifts trigger gratitude. Arrogant people typically think they deserve a great deal, and thus they take for granted many of their positive experiences. Such experiences are to be expected because of who they are, consequently they do not feel as much gratitude. To be sure, we should also be thankful for getting what we deserve. We do not create the situations that enable just deserts to be rewarded. The skills associated with gratitude include seeing many of our experiences as gifts and hence feeling thankfulness for them. We learn to focus on what others do for us, often beyond the call of duty. We become good at conveying gratitude in a genuine way that builds relationships. We learn how to approach even daunting situations with the tendency to see the gifts that lie therein. In times when conflict is commonplace, having well-honed skills of gratitude makes the world feel less cruel. Gratitude supports flourishing in most contexts, but its skillful habits are most needed when the times are turbulent and grim.

The hubris at play in our increasing political polarization over the last several decades has made issues of conviction and humility highly salient, and this has stimulated a vast interdisciplinary literature on humility.[25] We even see noteworthy academic centers devoted to issues surrounding humility, such as the Institute for Humility and Conviction in Public Life at the University of Connecticut. We have powerful cultural resources from which to draw lessons about cultivating humility, resources that should appeal to both conservatives and liberals. Humility is a core virtue in Christian traditions (and many other religions), and some of our iconic national leaders such as Washington and Lincoln exhibited key humility skills. But the allure of conviction, self-assertion, and status remains dominant. We need more concrete examples of how to build humility skills and to develop the reinforcing feedback loops that turn them into habits.

Building Humility Skills

We know from the last chapters that building the skillful habits associated with flourishing involves practicing them and finding mentors who can guide our practice. We also need to refine the feedback mechanisms involved in reflecting well on our performance. Reflection is not just a crucial element of humility; it is a necessary skill for cultivating the judgment required in any positive character trait. Friendships play an important role informing reflection and supporting practice, as is illustrated in the following stories. I begin with the story of Robert (Bob) Cabin's journey as a practitioner of ecological restoration, which he described in his book *Intelligent Tinkering*.[26]

In 1996, Bob Cabin was offered a postdoctoral fellowship at the National Tropical Botanical Garden, where he began working on restoration in the garden's Ka'upulehu Dry Forest Preserve on the Big Island of Hawaii. The preserve had been invaded by a very aggressive nonnative plant, African fountain grass, that was crowding out native species and creating a monoculture. Many of Hawaii's native ecosystems have been overwhelmed by invasive species and land-development pressures; as a result, the islands have more endangered and threatened species than any other state in the United States. What started as a two-year postdoc turned into a ten-year stint as a participant, planner, and leader of community-based dry forest restoration in numerous locations in Hawaii. During that time, Cabin wrestled constantly with the tension between designing projects to generate findings that would reflect and contribute to restoration science and designing them to maximize their ecological impact. He began his work with the conviction that good restoration should be conducted as science and should be guided by scientific knowledge. After his many years working in the field, his reflections led him to a much more pluralistic view of the practice of restoration, one that manifests the humility skills we have explored.

When he arrived in Hawaii, Cabin joined a small group of stakeholders who were deeply concerned about the fate of local dry forests. Started by a volunteer couple, the North Kona Dryland Forest Working Group had a great deal of passion but no clear leadership structure. Since Cabin had a doctoral degree, he was often treated as the leader, even though he had very little practical knowledge about restoration of these specific ecosystems. In such circumstances, some people would have enjoyed the attention and leveraged their degrees to turn their initial beliefs into convictions about how to proceed. But Cabin was a listener by nature. He was acutely aware of the knowledge he lacked and actively sought guidance from those who had been working in such ecosystems for years. He also dove into the rough work of pulling fountain grass and planting native replacements; he was not just a scientific adviser and researcher.

The tensions between restoration science and practice served as the impetus for Cabin's deep reflection, just as the tensions between our cultural conviction orientation and our current need for cognitive flexibility can motivate further development of our own humility skills. As a good scientist, Cabin created experimental plots to compare results of different treatments. He carefully monitored the plots and measured impacts of restoration treatments. Science should answer questions like whether pulling fountain grass is more effective than herbicide treatment, whether shade affects native seedling mortality rate, when one can stop watering seedlings, or which planting techniques most affect mortality rates. The grants he wrote aimed to inform future restoration by answering such questions. Most scientists would have measured their success largely in terms of knowledge acquired and publications, but Cabin was also concerned about how much actual restoration was accomplished. And, alas, the process of scientific restoration was both time-consuming and expensive.

After local successes, the Kona Dry Forest Working Group grew considerably, involving more agency personnel and more volunteers. It took on larger projects and used many more volunteers. The desire to scale up the group's activities contributed to deepening disagreements about a wide range of tactical questions. How much weight should be placed on the quality of the volunteer experience? How precisely documented should the treatments be? How much should the cultural significance of certain plants matter? Whose culture is most relevant? As Cabin scaled up his research, he needed to use more volunteers, but it was hard to rely on volunteers to accurately monitor the results of different treatments. Many volunteers participated because they wanted to make a significant difference rapidly, which led some crew leaders to adopt a "just get it done" approach to native planting or weeding. More was accomplished, but at the cost of good science.

Cabin's bosses wanted him to focus on the science, but the more he reflected on what the science was achieving, the less he was convinced that it was as important as maximizing restoration activity on the landscape. Over the course of the decade, he became increasingly concerned about the limitations of the knowledge generated by restoration science. Practical considerations, such as how to get water to new plantings and how many volunteers were available, dictated restoration planning as much as good science. Many of the issues he confronted had not been addressed in the literature, and where they had, Cabin found that often he could not generalize from other ecosystem contexts to his Hawaii projects. Minor differences in microclimate or plant interactions mattered more than ecosystem similarities. He became acutely aware of how little science can say about exactly how to restore a specific highly degraded area and increasingly sympathetic with the practitioners who said that their own trial-and-error application of practices that had worked elsewhere was often a more reliable guide than scientific papers. His own experience of developing elegant plans in the office and then finding them ill-suited to the idiosyncrasies of the site created an opportunity to learn from error that he grasped with relish.

The debates among personnel from agencies with different missions and volunteers with different values made working-group meetings longer and less helpful as the scale of operations increased. As a scientist, Cabin initially treated many of these debates as focused on what restoration strategies would be most effective, but increasingly he found they revolved around conflicting values, which science was not equipped to address. Instead of feeling like his expertise was increasing, he found himself more and more focused on the limitations of his training to address the issues that divided stakeholders. Indeed, he was curious about the diverse approaches to restoration represented in the group, and over time he became increasingly convinced that using multiple approaches provided a more reasonable way of responding to severe ecosystem degradation. There remained a tremendous need for more ecological restoration in the region, and it seemed foolish to alienate those who might help, even if the results were imperfect by scientific standards.

At the end of the book, Cabin recalls standing on the highway overlooking the area in which he started working in Hawaii, which was surrounded by huge tracts of highly degraded land, full of fountain grass. Despite the restoration achievements, their group had barely dented the problem. He had an epiphany. Instead of designing restorations in accord with best science, he concluded it would be better to promote a "meta-intelligent tinkering Adopt-an-Acre" program in which different groups of people with different goals would be able to pursue their vision for the land, within some agreed-upon constraints. Restoration would be better and faster if those in charge humbly acknowledged what they did not know and let lots of different projects proceed on the landscape rather than trying to impose a single standard on restoration projects.

The phrase "intelligent tinkering" comes from Aldo Leopold. Leopold was acutely aware of the limitations of our knowledge of natural systems and its implications for how we manipulate such systems. In his characteristically pithy fashion, he says, "If the biota, in the course of eons, has built something we like but do not understand, then who but a fool would discard seemingly useless parts? To keep every cog and wheel is the first precaution of intelligent tinkering."[27] Viewed narrowly, Leopold is urging us to preserve all species, even if we do not see how they contribute to system functioning. But Cabin extracts a larger message about humility from the idea that we should view our restoration efforts as a form of tinkering—that is, as a set of experiments where we try out different approaches to getting what we want without further compromising the larger system.

In this all-too-brief story about a decade of ecological restoration learning, we see the full set of reflective skills at work. Cabin begins by being aware of the limitations of his knowledge of Hawaiian ecology, but his awareness of limitations grows significantly to include limitations on the application of restoration science, on the capacity to generalize across similar ecosystems, on how to address value differences, and on how to manage group interactions. He begins with soft convictions about the role of science in doing good restoration, which he subsequently modifies both in response to the difficulty of applying them in the field and in order to work more collaboratively with others in the group. He consistently experiments and learns from his mistakes; his book is a chronicle of how he builds a network for peers to test his own reflective learning. But at the same time, he is also aware of the strengths he brought to the enterprise and proud of his accomplishments. As he puts the point, "Once again, I was proud of our research, the specific knowledge we gained and the more general academic contributions we made, and the indirect yet substantial ways in which our work contributed to the North Kona Dryland Restoration Working Group's restoration and outreach programs. Yet as valuable as all that science had been, I was painfully aware that we had not discovered much that was of direct practical value to the restoration of this and other degraded ecosystems."[28]

Cabin also exhibits the range of decentering-self skills. In the working group, he is more inclined to listen than to expound. He eschews the status and entitlements that might accompany his PhD and relishes the physical labor of restoration and the camaraderie that comes of sharing sweaty work. He demonstrates genuine curiosity about the motivations of volunteers who work for him and the experiences of practitioners who do not have advanced degrees. Throughout the book, he is effusive in his gratitude for all that people have contributed both to his development and to the work of restoring degraded landscapes. We can imagine other biologists reacting very differently to the experiences that Cabin has over these years—being more aloof, being more focused on maximizing publications, and being more convinced that science provides the answers about how to approach restoration. Cabin brings some humility skills to Hawaii, but these are reinforced and strengthened through the course of his work there. Throughout his book, we can see how his initial dispositions create positive feedbacks that refine and elaborate the skills in question. Someone bringing fewer humility skills would have been unlikely to arrive at a pluralistic intelligent tinkering approach to restoration and to become such a powerful reflective storyteller.

Bob Cabin's story reveals that the exercise of reflective skills is not a solitary venture. In our individualistic culture, we often emphasize the individual's primary responsibility for cultivating character through practice. Cabin's interactions with his restoration group play a crucial role in his growth. Others not only provide material on which to reflect and opportunities for practicing; they often provide direct feedback about how one is balancing competing values. Cabin developed friendships with people whose views he respected. As a result, he was open to learning from their feedback.

Here it is worth pausing to appreciate the role of developing a diversity of friendships in triangulating between humility and conviction and in augmenting other character strengths. This idea has a long history. Aristotle believed that friendship was crucial for developing virtue for two main reasons.[29] First, having genuine friends brings out the best in people. We care for friends' welfare for their own sakes, and thus we work hard to act in ways that promote their interests even when these are in tension with our own narrow self-interest. Second, friends will challenge us when we fall short of the character to which we aspire, because they want the best for us.[30] Both of these points figure in helping us to triangulate between conviction and humility. In learning how to be a friend well, we must learn how to navigate conflicting convictions, often softening them to enable concord. But we must also support conviction, when forceful action is required. Just knowing someone cares and believes in you often provides the support that enables action in uncertainty.

Rachel Carson's deep friendship with Dorothy Freeman seems to have played a crucial role in providing such support for the process of writing *Silent Spring*. Carson wrote a letter to Freeman that included the following tribute to this friendship:

> I don't suppose anyone really knows how a creative writer works (he or she least of all, perhaps!) or what sort of nourishment his spirit must have. All I am certain of is this; that it is quite necessary for me to know that there is someone who is deeply devoted to me as a person, and who also has the capacity and the depth of understanding to share, vicariously, the sometimes-crushing burden of creative effort, recognizing the heartache, the great weariness of mind and body, the occasional black despair it may involve—someone who cherishes me and what I am trying to create.[31]

For both Carson and Aristotle, such genuine friendships are rare, but that depth of friendship is not required for the aid to reflection that friends can provide. If we cultivate relationships of mutual respect and care that are characterized by honesty and a diversity of viewpoints, we will find both support and challenge that strengthen our reflective skills, providing we are open to the feedback.

Like other character traits, humility may be expressed in some situations and not others. We may have a set of skillful habits that we do not use consistently. For example, it could be a mistake to generalize from Cabin's use of humility skills in Hawaiian restoration to his having that broad character trait. Indeed, friends often help us to see such inconsistencies in our behavior. But learning skills in one context often does inform behavior in other contexts. Wondering whether Cabin had a hard conviction that intelligent tinkering was the best approach to restoration across contexts, I asked him how far he would extend his pluralism. He replied that he has become more and more skeptical about generalizations of many sorts, including about whether intelligent tinkering itself should be generalized to all restoration. What began as a recognition that ecosystem variability makes generalization in ecology risky expanded to include most generalizations about restoration and about science's role in policy.[32] In his case, the humility skills were applied increasingly broadly, as he reflected on the importance of the lessons he learned.

Cabin's story illustrates how opportunities to use our expertise in leadership roles provide numerous occasions for learning how to better integrate conviction and humility. In such circumstances, questions about how much we know collide with questions about what we need to do to lead effectively. On which of these should we focus? How can we integrate both into a kind of binocular vision that leads to intelligent tinkering solutions? I suspect that the puzzle with which we started this chapter reveals an inevitable tension in life between what people reasonably seek in leaders and ways we should all approach empirical questions. We can learn to reduce this tension by developing sophisticated humility skills, but we cannot eliminate it. However, if we do succeed in shifting social norms toward greater emphasis on humility, then the demands on leaders to express hard convictions will diminish and the tension will be reduced.

Both humility and collaborative skills are important for flourishing now, in large part because they strengthen our group problem-solving capacities, reduce social polarization and conflict, help us to broaden our understanding of the systems on which we depend, and help us to rapidly build new relationships with diverse peoples. We have seen that these benefits are particularly important given the challenges and opportunities characteristic of the age of climate change. We now turn to frugality skillful habits, which engage very different aspects of our lives—how we consume, what we find valuable, and how we relate to nonhuman nature. We will find though that all three clusters of skillful habits are mutually reinforcing. Humility is often associated with frugality. In the Christian Gospels, both are linked to core virtues exemplified by Jesus. In Taoist traditions, the pair, along with compassion, are the "three treasures" or core virtues. We often find that those who are unconcerned with status are also unconcerned about material possessions. Those temperate about their convictions are also temperate about their consumption.

4

TECHNOLOGICAL FANTASY AND FRUGALITY

On September 29, 1980, Paul Ehrlich and Julian Simon finalized the most famous bet associated with the environmental movement.[1] They bet about whether the price of five metals would go up or down over the next decade (tin, nickel, copper, chromium, and tungsten). Ehrlich reasoned that increased population pressure and resource use would drive up the inflation-adjusted price of the metals, and Simon maintained that the free market and technological development would create enough abundance that prices were likely to go down. It requires a bit of background to see how the bet reflected two opposing views about our overall impact on the environment and human welfare.

In 1968 Ehrlich, a prominent Stanford University biologist, wrote an alarming book, *The Population Bomb*. He argued that human population was rapidly outstripping the resources necessary for humans to flourish and that mass famine and death would ensue even with serious efforts to reduce population growth. Thomas Malthus had voiced similar concerns as early as the 1790s. But the specter of near-term human population overshoot was highly salient in the 1960s. Another book, *Limits to Growth*, argued that unless we changed course, by the middle of the twenty-first century we would overshoot the capacity of the earth to provide the resources necessary to sustain our industrial society.[2]

By contrast, a group of futurists who are often characterized as Cornucopians believed that the free market and technological development would create material abundance even with rapid population growth. Simon, an economics and business professor, argued that as demand increases for nonrenewable materials like copper, we are unlikely to run out of them, because their prices will go up, and that will increase the motivation for recycling the materials, using them more efficiently, and finding alternatives. As a result, eventually the price of metals should go down as technological development increases the supply of alternatives. On this view, creating feedback loops that stimulate human ingenuity and fuel technological development is the best solution to concerns about growth, not restraining the growth of population or consumption. Thus, the bet was an expression of a deep conflict between ideological frameworks.

Simon won the bet. By 1996, the price of all five of the metals had decreased in inflation-adjusted dollars. Ehrlich paid him $576 dollars, which was the total decrease of $200 worth of each metal over the decade. Ehrlich did point out that if they had selected a different decade, the results would have gone the other way. No simple bet could resolve the conflict between these ideologies. But this result, and many like it, did give a boost to Cornucopian views and the skillful habits they emphasized during the last part of the twentieth century. I will call those habits "abundance skills" because their use is supposed to lead to the material abundance that Simon envisioned. The contrasting skillful habits associated with Ehrlich's view are aspects of frugality, broadly understood: they involve "doing more with less" and thereby reducing the human impact on the natural world.

In eighteenth-century America, frugality was an ascendant social norm; people were proud of the skills that reflected the norm. Homesteaders in the New World engaged in frugal practices out of necessity. Benjamin Franklin celebrated such norms in his *Poor Richard's Almanac*. But by the middle of the twentieth century, abundance norms had become dominant in the culture; the skills associated with the acquisition of material wealth had become central to a version of the American dream.

Four key cultural norms have expressed our current approach to material abundance. First, success involves financial wealth and material prosperity. Whatever else we pursue, if we achieve the status associated with success in our culture, we will have the wealth that enables us to enjoy conveniences, travel, fine foods, and so on. Second, technological development is progress; to improve human welfare and to address major issues, we should focus on increasing the sophistication of our technologies.[3] Third, free markets typically produce and allocate goods in ways that enhance human welfare. Government may assist markets with subsidies, expertise, and safety nets, but markets far more efficiently allocate goods and build wealth. And fourth, growth is good. In general, the larger some group is, the more successful and desirable it is. Smaller organizations may be more nimble, but their goal is typically to grow their assets and become larger.

Much has been achieved as a result of our emphasis on material abundance norms and skills. We have modified much of the earth's surface and turned what we have found into commodities that make our lives easier. Our improved agricultural yields enabled us to avoid the famines that would otherwise have accompanied our population growth. Our advances in medicine keep us alive much longer on average than prior humans. We have advanced our understanding of the world so we can reliably predict how to satisfy our growing desires. We have achieved average levels of prosperity that would be incomprehensible to our ancestors. Indeed, we have become one of the most powerful forces on earth.

And yet, we know that our increased material standard of living is a mixed blessing. Increasingly its negative side effects are more salient and more dangerous. Support for abundance norms has eroded in some quarters. Polling data shows steadily weakening support for capitalism among young people, and concern about the impact of technology regularly surfaces.[4] "Small is beautiful" has been a slogan for some for over fifty years and has been renewed by the relocalization movement. Increasing resistance to abundance norms provides an opening for a revitalization of the frugality norms and skills I describe below. But material abundance norms still explain a great deal of our collective behavior and our institutions. For example, our dominant measure of progress is GDP, which favors financial growth. Our legal system makes it very hard to slow technological development or to constrain technology that does not have obvious deleterious consequences. Our schooling increasingly emphasizes job skills for the new economy.

Our abundance orientation is problematic now for three related reasons: First, as we saw in chapter 1, we are reaching planetary boundaries that technology alone cannot address, thus it seems very doubtful that we can sustain continued growth in population and consumption. Second, it is increasingly clear that material abundance and its technological engine are not really increasing human welfare; more material abundance seems unlikely to enhance flourishing on average in wealthy countries. And third, we have been able to sustain our material abundance so far only because it is not distributed justly across the population. The unfairness of this distribution has become more salient, and increasingly people are unwilling to accept the injustice. As we unpack these arguments, we will see why a rebalancing toward frugality will increase our resilience in the short term, and if widely accepted, it will facilitate a movement toward sustainability in the longer term.

Frugality, Necessity, and Self-Restraint

For many people, frugality or thrift seems an outdated virtue. Yates and Hunter note that in the last fifty years of the twentieth century, "thrift seemed to go the way of chastity, teetotalism and other quirky artifacts from a more morally uptight age."[5] Others think it *ought* to be outdated, for it suggests a small, crabbed existence of the sort manifested in T. S. Eliot's Prufrock, who has measured out his life in coffee spoons. Hart says, "As a value thrift is devoid of any element of the majestic or the mysterious, and so is impotent to stir our imaginations or inspire our wills; it does not enchant, beguile or inveigle us."[6] In an age when character is regaining significance in public discussion, the major psychological work on the topic, Peterson and Seligman's *Character Strengths and Virtues*, does not even mention frugality in its eight hundred pages.[7] To be sure, the simple-living movement continues to have some attraction, and environmental critiques of consumption are common, but few people turn to the historically important virtue of frugality to address the consumption issues that are at the root of our unsustainable lifestyles. Why is this so?

Part of the answer, I suspect, lies in its historical connection with necessity and deprivation. Poor peoples around the world typically exhibit frugality with great skill because they have to. Frugality played a key role in early American thought partly because of the scarcity of consumer goods and partly because of the influence of Puritanism and Quakerism.[8] Most early settlers worked the land with tools they made and saved any surplus they had for hard times. But by the time Ben Franklin popularized frugality, consumer goods from England were more readily available in cities, and it was necessary to identify other motivations for thrift.

A similar pattern occurred during the Great Depression of the 1930s, when a generation found it necessary to cultivate skills that early Americans largely took for granted. The generation that followed would either learn the skills without the need or abandon the virtue. Some young people today still see college as a time of enforced frugality, usually to be abandoned afterward when they reap the rewards of the degree. Sometimes nostalgia for a simpler time, when survival determined one's priorities, extends frugality beyond necessity, but it tends to be a weak motivation in the face of the lure of luxury.

Another partial explanation of frugality's unattractiveness is that we have an overly narrow conception of frugality. Like collaboration and humility, frugality is uncharitably understood today. The rise of consumerism as an engine of economic growth has led us to associate frugality primarily with a willful version of self-restraint. We have more and more opportunity to consume a vast variety of goods and services, and we are surrounded by advertising. It just takes one click on Amazon to satisfy a desire induced by some advertiser. Thus, we find that avoiding consumption requires increasing self-restraint, which can be exhausting. Self-restraint *is* important for one kind of frugality, which we might call restrictive frugality. However, we need to recover a broader conception of frugality—what I call constructive frugality, which is much more attractive as a core character trait for our times. Constructive frugality aims at building a life rich in intangible goods such as friendship, community, beauty, and spiritual growth and de-emphasizes the monetary and material spheres of life.

Let us begin with the narrow conception and build toward constructive frugality. Restrictive frugality includes skillful habits that enable us to avoid spending money and producing waste; many of these habits involve strategies of self-restraint. We are tempted to buy now (and pay later), to seek convenience, and to find status in possessions, but restrictive-frugality skills involve deferring gratification and making do in order to live well later.

The example of Benjamin Franklin's early life and writings provides a useful window into some of the skills of restrictive frugality.[9] Franklin was born in Boston in 1706, one of seventeen children fathered by Josiah Franklin, a tallow chandler. He was a bright boy and was sent to grammar school with the idea that he might join the clergy. But his father ran out of money, and Ben was apprenticed to his brother, who was a printer. After learning that trade, but coming into conflict with his brother, he ran away, first to New York and then to Philadelphia at the age of seventeen. Penniless, he found work in some printers' shops, but he yearned to open his own printing business. After an unsuccessful trip to London to secure a printing press and time working off his trip as a clerk and bookkeeper, at twenty-two he entered into a partnership with Hugh Meredith to start a printing house. In these lean times, his frugality was a necessity.

While some successful merchants of the times flaunted their wealth, Franklin deliberately crafted his persona in frugal terms. Famously, he carried his paper in a wheelbarrow through the streets of Philadelphia. He dressed plainly and would not use china at his table. Because he had taken on considerable debt to start his business, he was as concerned to appear frugal as to live in accord with frugality's dictates. His common-law wife Deborah Reed was similarly inclined. In his autobiography, Franklin says of this time, "It was lucky for me that I had one [a wife] as much dispos'd to Industry and Frugality as my self. She assisted me cheerfully in my Business, folding and stitching Pamphlets, tending Shop, purchasing old Linen Rags for the Paper-makers, &c. &c [etc.]. We kept no idle Servants, our Table was plain and simple, our Furniture of the cheapest. For instance my Breakfast was a long time Bread and Milk (no Tea), and I ate it out of a twopenny earthen Porringer with a Pewter Spoon."[10]

Franklin started publishing a newspaper, the *Pennsylvania Gazette*, when he was twenty-three, and before he was thirty he started to write and publish the immensely popular *Poor Richard's Almanac*. Issued annually for twenty-six years, the *Almanac* sold more than ten thousand copies each year and significantly enhanced Franklin's income and reputation, even though he authored it under the assumed name Richard Saunders. Many of Poor Richard's proverbs about frugality remain in circulation today:

> Beware of little expenses, a small leak will sink a great ship.

> Buy what thou has no need of and ere long thou shall sell thy necessaries.

> All things are cheap to the saving, dear to the wasteful.

> Who is strong? He that can conquer his bad habits. Who is rich? He that rejoices in his portion.

> A penny saved is a penny got.[11]

These proverbs emphasize dispositions to save, to restrain consumption, and to avoid waste, which are central themes of restrictive frugality, but they do not highlight the skills that strengthen the dispositions. Franklin's early lifestyle reveals many of these skills. He had mechanical skills that enabled him to fix broken machinery. He often found new uses for worn items rather than discarding them. He developed a successful strategy of systematic self-evaluation for reinforcing the thirteen virtues he wanted to perfect, including frugality. He also engaged his social skills in founding the Junto, a club that aimed at mutual self-improvement, which met on Friday evenings to discuss questions that Franklin developed. At Franklin's suggestion, the Junto created the first lending library in the United States, in part as a way to avoid buying books that might be more economically shared, thereby reducing material consumption while expanding access to scarce material goods. Franklin enjoyed significant wealth and the luxuries that it enabled later in life, but as he tells the story in his autobiography, his early self-restraint in the face of temptations to consume led to the wealth he acquired.[12]

It is useful to separate self-restraint skills into several categories that can be learned in different ways. These skills are well known, but they often are underdeveloped where necessity does not require them. The first group involves techniques for managing temptation; it includes strengthening willpower, distracting oneself from a salient desire, shifting focus toward a longer-term goal, and avoiding circumstances where temptations will be powerful. Walter Mischel's famous "marshmallow experiments" showed variations in young children's acquisition of such skills. Mischel tested four-year-olds' abilities to delay gratification.[13] Children were presented with one marshmallow and told that they could eat it, but that if they waited for fifteen minutes before eating it, they would receive a second marshmallow. Other parts of the experimental conditions were varied to see how they affected the results. Children who successfully delayed gratification to get the second marshmallow often used creative ways of distracting themselves. They turned away from the marshmallow, they sang songs, fiddled with their hands and feet, or invented games to pass the time. Such skills reduce the need for direct use of willpower. Follow-up research showed that those who were better at delaying gratification tended to do better in school and in social interactions more than a decade afterward.[14] Many other studies have shown links between the broader category of self-regulation skills and achieving positive outcomes.[15] The unpopularity of frugality may result in part from the focus on pure willpower, which in a consumption-oriented society may be exhausting over time. However, as the marshmallow experiments demonstrated, self-restraint skills include distraction skills, whose use is much less exhausting.

A different set of self-restraint skills involves making do with what one has—focusing on a larger goal and finding ways to achieve it without additional material consumption. Franklin's creation of a lending library now seems an obvious move, but it was quite creative at the time. His use of a wheelbarrow and human power to move his paper stock enabled him to avoid the expense of a horse-drawn carriage. Examples from the twenty-first century include my replacing the battery on an older cell phone rather than buying a new phone, or my friend's use of old bed frames as a trellis for her peas rather than buying a new trellis. Often, repurposing an object to serve another need—for example, using old bottles as a translucent building material—creatively addresses numerous goals.

The last kind of self-restraint skills I will mention are satiation skills, which typically involve accurate self-monitoring, a fine tuning one's sense of enough, and finding the pleasure in becoming a person who stops consuming when approaching satiation. Franklin's systematic self-evaluation served to put brakes on his pursuit of desire satisfaction and in effect strengthened his satiation skills. We know that some people habitually overindulge in food and drink because of the immediate pleasure it gives; they have relatively weak satiation skills. On the other end of the continuum are people who take great pleasure in self-denial; they are habitual ascetics. Ascetics rarely approach satiation and arguably live less fully as a result. People with strong satiation skills occupy the middle of this continuum. They pursue what they desire and live fully, but they have good brakes on their consumption.

Restrictive frugality's emphasis on self-restraint evokes Plato's image of our needing to pull hard on the reins of the steeds that represent the passions, but we have just seen that a number of the skills involve less willpower and more creativity. With such diverse skills, the work of curbing desire satisfaction need not be exhausting. A similar creativity is present in the next cluster of frugality skills—those in which consumption of market goods is restrained through the use of self-provisioning, repair, and reuse (SRR) skills. These skills involve self-restraint in the broadest sense: they reflect a tightening of the belt when it comes to material consumption. Thus, I lump them in restrictive frugality, but they do not reflect the direct use of willpower to avoid consumption. They serve as a bridge to constructive frugality.

The broad class of SRR skills is often emphasized in the simple-living movement, where people grow their own food, forage or hunt for wild foods, preserve foods for winter and save seeds for spring, build their dwellings, heat with wood, and fix older goods rather than buying new. The specific skills involved are too numerous to describe in any detail and too well known to require such detail. They are often described in ways that suggest they represent a version of the back-to-the-land movement in which people move out into the countryside to homestead on a small patch of land. Indeed, one historically important source for these skills is the Foxfire series of books, which gathered stories and practices of rural living in Appalachia and played an important role in guiding the 1960s and '70s' back-to-the-land movement. With the rise of urban gardens, home cooking, foraging in parks, and living in tiny houses, such skills now apply much more broadly. Kaplan and Blume's urban homesteading book sketches many of these adaptations for urban frugal living.[16]

A common worry about developing SRR skills is that their use is time-consuming. It would be more convenient to just buy what we want. But other benefits arise from their use that more than offset the time demand. In *Braiding Sweetgrass*, Robin Kimmerer writes eloquently about the knowledge and the joy that come from self-provisioning interpreted through her Native American heritage.[17] Gardening connects us to the soil and teaches us the ways of plants. Gathering wild strawberries engenders appreciation for the gifts that nature bestows on us. Over time, we can learn how to integrate ourselves beneficially into our ecological communities and develop reciprocal relations with our more than human neighbors. The beauty and wisdom that lie therein lead us into the territory of constructive frugality.

We often associate SRR skills with increasing self-sufficiency. Instead of depending on complex supply chains to supply our food or professionals to repair our household goods, we learn how to rely more on ourselves for these necessities. But this picture is somewhat misleading, since few can master the wide range of skills necessary for individual self-sufficiency without sacrificing many of the advances of modern society. I know how to grow vegetables; my partner is skilled at making cheese and canning foods. I rely on a friend to help improve my composting. My neighbor across the street knows how to weld broken tools, and my brother can repair common machinery and fix electrical problems. We inevitably depend on a community of people who have different SRR skills, thus self-sufficiency is more easily approached at the community level than the individual level.

But can we realistically acquire enough of these skills? Many products like cars, cell phones, and computers are made in ways that render repairs so complex that only a professional can do the job, or so costly that buying new seems more efficient. Moreover, SRR skills are rarely part of standardized K–12 schooling.[18] People who want to acquire such skills typically learn from family members or social networks, but in our increasingly postindustrial world the opportunities for learning many SRR skills appear to be declining. For those who cannot find community members to teach SRR skills, the internet provides a tremendous variety of sources that fill the gap, including YouTube repair videos, the giant Ifixit website (www.ifixit.com), and blogs on how to grow vegetables, hunt, and gather wild foods. The motivated can learn a great deal.

Box 3 summarizes the clusters of frugality skills we have surveyed so far and those associated with constructive frugality, to which we now turn.

Box 3. Frugality skills
Self-restraint skills

- Managing temptation and deferring gratification
- Making do
- Satiation skills—honing one's sense of enough

Self-provisioning, repair, and reuse skills

- Growing food, hunting, and fishing
- Basic carpentry, plumbing, and mechanical repair skills
- Repurposing skills—finding new uses for old items

Constructive frugality skills

- Seeing the beauty in imperfection
- Finding social, spiritual, and cognitive fulfillment
- Touching the world lightly

Both self-restraint and SRR skillful habits are challenging to motivate when prudence does not require them and where material consumption is central to a good life. To more powerfully motivate cultivating frugality, we must turn to constructive frugality, which absorbs the practices associated with restrictive frugality into a life focused on pursuing social, aesthetic, moral, and spiritual good.

Reenvisioning Frugality

Simple-living movements, from Epicureans and Stoics to Buddhists and Christian monks, tend to be expressions of constructive frugality. Insofar as the Puritan and Quaker sources of frugality in early America aimed at glorifying God or on strengthening community relations, they too would primarily motivate frugality through a shift in focus to nonmaterial goods. The simple-living movements that have periodically appeared as a reaction to too much worldliness typically involve this approach to frugality. Bouckaert, Opdebeeck, and Zsolnai's *Frugality: Rebalancing Material and Spiritual Values in Economic Life* contains numerous articles that highlight the benefits of constructive frugality in the current age.[19] They analyze the impacts of the worldliness of consumer culture and recommend a rebalancing toward frugality both in business and in how we live. Juliet Schor argues that we should pursue reduced work hours, accompanied with shrinking incomes, and allocate more time to provisioning and sharing with neighbors. She sees this trend as a way of finding plenitude rather than decreasing our quality of life.[20]

Many of the skills associated with restrictive frugality are also central to constructive frugality. Being able to make do with what one already has is central to both; repurposing goods, repairing what is broken, creatively reusing waste, and sharing with others remain at the core of constructive frugality. But a distinct cluster of skills is highlighted by constructive frugality, including being good at finding beauty wherever one is, building social bonds and enjoying their fruits, and finding opportunities for growth where others find only travail. Epicurus is an excellent guide to some of the practices associated with constructive frugality. We learn to distinguish pleasures that are typically followed by pain from those that are not, and we pursue the latter. We habituate ourselves to a simple, nutritious diet. We pursue friendships and detach from worldly goods.[21] While some techniques of self-restraint are still important here, they are not the focal skills, since one no longer craves that which once required restraint to avoid. We now look at three of these distinctive constructive frugality skills in more detail. The first is finding beauty in our daily lives.

Some experiences of beauty require significant expense—for example, traveling to the Swiss Alps, purchasing elaborate interior décor, or dining at a three-star Michelin restaurant. In the material abundance paradigm, such experiences set the standard for intense beauty, and only the wealthy have regular access to them. If, however, we broaden our repertoire of experiences that provide intense beauty, we find that many of these are virtually free. Moreover, as we learn to notice the beauty that is available without expense all around us, the pleasure it provides reduces our need to consume material goods. This is especially evident in the beauty that characterizes many of our natural and built environments.

Edward Abbey, the great environmental author, once said "there is beauty, heartbreaking beauty, everywhere."[22] He is musing about why he loves the desert so much, but he maintains that it is not just beauty, because one can find beauty in many natural and human settings, even in "the back alleys of Hoboken, New York City, Berlin . . . and Pittsburgh." Abbey is *not* saying everything is beautiful. Without contrasting ugliness, finding beauty would be meaningless. He is also *not* just urging us to see the good—the beauty—in every place. This staple of positive psychology is too often interpreted as recommending an uncritical, Pollyanna-style approach to happiness, which Abbey eschews. Abbey is suggesting that we can learn how to find genuine beauty even in austere, flat, dusty deserts like the Pinacate in Mexico; and not just in their stunning sunsets and spring wildflowers. But what does this learning involve?

Most of us have learned how to appreciate culturally standard forms of beauty. Without effort, we see certain people as beautiful, we appreciate the brilliant colors of autumn leaves, and we hear the resonant power of certain chord sequences. But constructive frugality requires moving beyond these perceptions to experience a much broader range of beauty, such as the beauty in many older people, the beauty in city back alleys, and the beauty in struggling to master some sport. The skill of finding beauty where it is not obvious often involves learning about its history or its components. We may need to learn about its role in larger systems and the meanings that others give to it. Such knowledge enables us to see relationships that are not obvious. Terry Tempest Williams describes the threatened Utah prairie dog's life in ways that help us to appreciate the beauty of the animal and the system it creates.[23] Such knowledge helps us hone our ability to notice obscure details in a setting that reveals its beauty. A dissonant jazz piece may be heard as beautiful, but only if one knows what the musicians are attempting and notices the patterns they weave. It can take much practice to experience such patterns. Similar points can be made about the song of the robin.

Cultivating a broad sense of beauty also involves connecting what we notice to poignant emotions. Take the beauty of Edward Hopper's *Nighthawks*, a painting of three customers sitting in a late-night diner whose glow illuminates the empty streets. Its beauty comes in part from the painting's powerful evocation of loneliness. The loneliness is not itself beautiful, but the poignance of the painting is. What a painting can do, our own experiences of street scenes can also do. Abbey's appreciation of the desert has some of this magic.

I have found tremendous joy in expanding my own appreciation of beauty on our farm. When chores call, it is so easy to quickly walk past spiderwebs glistening with dew, yet just pausing there for a minute reopens me to my surroundings. For years, I walked past the moths that settled next to our windows, without appreciating their stunning array of subtle colors. Now I photograph them and look them up. Buying a cheap set of watercolors and trying to paint landscapes on long winter evenings taught me to appreciate the magical complexity of clouds, which I do on a daily basis. Asking ourselves how to paint something is a sure-fire way to focus on its subtle patterns. With practice, I even learned to see the brief flashes of color in the cedar waxwings chasing each other over the pond. The beauty repertoire I have learned to expand is mostly in nature, but that has inspired me to find the beauty in other unusual places.

A second group of constructive frugality skills functions similarly to finding beauty but expands the kinds of satisfaction that we can experience with minimal material consumption in our daily lives. These skillful habits enable us to increase our social, spiritual, and cognitive fulfillment. If we improve our capacity to acquire fulfillment from social, spiritual, and cognitive dimensions of life, we tend to focus less energy on material consumption. Of course, such fulfillment often involves some material consumption, as would strengthening social bonds by inviting people to dinner, but because material goods are only means, not the ends of the activities, their importance is greatly reduced.

Habitually finding fulfillment in social, spiritual, and cognitive dimensions of life involves a broad range of skills. People typically excel at only one or two of these dimensions. The spiritually skillful often have learned how to find meaning in ordinary struggles that would otherwise be tedious and oppressive. They often know how to achieve equanimity amid strife through meditation. They know how to call forth joy in response to ordinary events and compassion for many forms of suffering. The socially skillful typically have the collaborative skills sketched in chapter 3, but they are also good at building friendships, making acquaintances feel valued, and organizing events that build community. Those skilled at finding fulfillment through cognitive endeavors such as learning a new language or a complex game, mastering ArcGIS or bird identification, typically have the curiosity skills discussed in chapter 3 and also research skills, memorization skills, and the skills of making fine distinctions. The more developed such skill sets are, the more likely one will be to find pleasure in using them. Such pleasure is the motivational engine for constructive frugality.

The third constructive frugality skill set is more general than those above and less familiar in the West. It is best described metaphorically as "touching the world lightly." Like the material abundance orientation, the pursuit of intangible goods can be pursed in a heavy-handed fashion that emphasizes increasing acquisition. The aggressive consumption of nonmaterial goods—acquiring a vast quantity of "friends" or beautiful digital pictures—seems antithetical to the spirit of frugality. Also, the dominance of intangible goods can cut us off from the values of material existence, like the taste of an excellent craft beer or a well-cooked meal. The skills of touching the world lightly involve a habitual moderation that enables both saving and savoring things of value. They integrate both self-restraint skills and a balanced skillful engagement with intangible goods into a form of artful living, which is good in itself.[24] Such skills include honing a keen sense of what is enough in each realm of life, acquiring what one needs without force, and using the least effort to secure the goods we desire.

If we look for traditions that seem to cultivate touching the world lightly, Taoism stands out. Frugality is one of the three treasures identified in the *Tao Te Ching*: "Given frugality, I can abound."[25] Its central position as a virtue seems clearly associated with Taoism's pervasive theme that when one aligns one's action with the *Tao*, benefits flow without effort. The character trait expressed by 儉 (*jian*, or frugality) is not just a means to some other end but rather a beautiful way of being in the world—an end in itself. To be sure, this skill of acting by touching the world lightly is said to create abundance, but that comes as a byproduct, not the goal. The goal expressed partly in terms of frugality is aligning one's virtue with nature. Taoism's central metaphor of acting like water involves not grasping for more than one needs and acting without heaviness—without exerting effort.

The skills involved in touching the world lightly enable a non-acquisitive, non-consumerist approach to financial, material, relational, aesthetic, and spiritual dimensions of life. They undergird the ability to enjoy without overindulging and the judgment to know in a context when to avoid temptation and when to savor something of value. They permit developing a taste for excellence in music, or food, or clothing, but while exercising such taste sparingly. Their creative use yields a motivationally powerful artful approach to life, which celebrates intangible goods while appreciating the value of material goods. If frugality is seen as involving the skills of touching the world lightly, it seems capable of being a central organizing trait for our lives amid the turbulence of our times.

Frugality and Flourishing

Although constructive frugality skills are more intuitively attractive than reducing material consumption through self-restraint, their appeal is likely to be insufficient for many embedded in the material abundance paradigm. If we can afford the convenience and pleasures associated with material consumption, why not pursue them? Emrys Westacott's fine book *The Wisdom of Frugality* summarizes many reasons for cultivating frugality.[26] I will focus on those reasons that are tightly linked to the age of climate change. Westacott notes that the distinction between prudential and moral reasons for being frugal is often very blurry. We often have many reasons for cultivating a character trait, and frugal skills easily serve multiple goals. Still, the distinction helps us to organize the case for frugality. The prudential reasons that begin our summary may convince the broadest range of people, though the moral reasons may make the most powerful case.

Perhaps the most salient prudential reason for cultivating frugality skills is that they enable us to reduce spending and thereby to save our funds to tide us over in hard times. The increasing turbulence we should expect in the age of climate change makes job losses, recessions, social crises, and natural disasters more common. Having some financial buffer increases our resilience and reduces economic anxiety. This reasoning may not induce the wealthy to cut consumption much, since they may already have adequate financial buffers, but it provides good reasons for most of us. Moreover, these skills also enable more saving for investment in new enterprises that may grow during reorganization phases of the adaptive cycle. During rapid growth phases of the adaptive cycle, lack of adequate investment can leave promising initiatives under-resourced, which increases their failure rate. Fath, Dean, and Katzmair call this "the poverty trap." Increased saving would also reduce the economic anxiety that many Americans have about retiring comfortably and would reduce the emotional burden that often accompanies large student loans and credit card debts.[27]

The above financial buffer argument for frugality is complemented by the physical buffer that self-provisioning, reusing, and repurposing skills can provide in crises. As turbulence increases, supply chains are more likely to be disrupted, making it hard to acquire goods that meet basic needs, like food, transportation, heat, and clothing. Those who know how to make do or self-provision are more likely to meet their needs amid such disruptions. It helps if our needs are fairly simple in the first place.

Those whose constructive frugality involves cultivation of a wide range of social skills are likely to have social buffers that augment financial and physical buffers in crises. A more densely woven social fabric increases the general resilience of community members.[28] It strengthens generalized trust, and thus we need not allocate as much time to protecting against those who might take advantage of us. The resilience one acquires from these buffers is not just valuable in severe crises: it provides peace of mind when disruptions threaten, which is all too common. We are more likely to flourish when our anxiety about meeting needs is low. Those living in poverty often cannot develop much financial buffer, but they often have strong extended families and social networks that offset some of their financial disadvantages.

Another prudential reason for cultivating frugality skills highlights the comparative impact of material abundance and constructive frugality on human happiness. Survey data indicate money and material goods provide limited increases in happiness. Between 1962 and 2018, US personal consumption spending increased from $2.15 trillion to $13.3 trillion in inflation-adjusted dollars.[29] The mean square footage of new single-family houses increased by 60 percent between 1973 and 2015.[30] Our information-processing capacity as measured by operations per second doubled every eighteen months between 1971 and 2009.[31] Our communications and transportation technologies have dramatically expanded our opportunities. We have a great deal more stuff and more opportunities on average, yet according to the US General Social Survey, our overall happiness has been roughly static across this period, and it has dropped somewhat since 2000.[32]

The doubtful connection between material abundance and well-being is reinforced by two other kinds of studies. A meta-analysis of 259 studies found a significant negative correlation between materialist values and well-being.[33] Here, having materialist values involves focusing on acquiring money and possessions that indicate status, which is only one aspect of the material abundance orientation. This meta-analysis measured the values people have, not their actual possessions, so it does not show that *having* money and possessions is negatively correlated with well-being. But if valuing something highly tends to makes us less happy, then we should consider emphasizing other sources of value.

Another kind of study looks at correlations between level of income and measures of well-being. A number of studies have indicated that below incomes of approximately $75,000 per year, increased income is positively correlated with increases in happiness as measured by balance of positive and negative affect; but above that income level there is no such correlation. Struggling economically is tough on happiness, but getting a big raise once one is already doing well financially is not likely to increase happiness over the long run. My argument here is not that material abundance has no relation to human flourishing, but rather that in our context, there is considerable reason to doubt that doubling down on material abundance norms and skills will be the best way to increase flourishing in the United States and other developed nations.

By contrast, strong social relationships, the enjoyment of beauty, and spiritual goods seem to yield deeper and more long-lasting satisfaction. Perhaps the most robust finding in happiness research is that having high-quality social relations correlates with well-being.[34] This applies both to close relationships and more distant friendships. People recover from illnesses faster if they have strong relationships; they suffer less loneliness, and they feel greater satisfaction with their lives. Robert Waldinger, director of the Harvard Study of Adult Development, puts the point succinctly: "The lesson that came from tens of thousands of pages of that research was that good relationships keep us happier and healthier."[35] Spiritual practice and finding beauty throughout our surroundings also seem to be strongly correlated with happiness. Today many find that engaging in spiritual practices and seeking beauty in nature lead fairly reliably to an inner peace, which Epicurus argued is the highest form of pleasure.

The happiness benefits of frugality, however, are not just a function of pursuing intangible goods. The more frugal we are, the fewer hours we need to work at a job, and the earlier we can retire. Sometimes a job is a joy, but most people would prefer more work/play balance. Working less enables us to spend more time pursuing other activities that enhance our lives and provides more time to practice self-provisioning, repair, and reuse skills. Juliet Schor makes the shift to working less a cornerstone of her low-consumption "plenitude" economy.[36]

The above prudential reasons for cultivating frugality tend to support our individual flourishing no matter what anyone else does. The following moral reasons are most powerful if they influence the actions of many people, since they focus on the environmental and social impacts of material consumption. By reducing our material consumption, we decrease our contribution to the environmental degradation associated with creating and disposing of physical products. We also reduce our contribution to consumption injustices resulting from economic inequality.

Most environmentalists have long ago internalized the IPAT formula—that negative environmental impact (I) is a function of population (P), affluence or consumption (A), and technology (T), which can mitigate or accelerate negative impact. Having just one child (or none) is one way to reduce impact, in part by reducing future consumption.[37] We can use technologies that reduce impact (e.g., solar panels and electric vehicles). But one of the most direct ways to reduce impact for people in relative wealthy countries is to reduce consumption and/or to shift to less impactful consumption. A growing "conscious consumption" movement is shifting product marketing and presumably business practices, but this does not deal with the root problem of overconsumption. As John Ehrenfeld puts the point, "Most of what businesses are now doing in the name of sustainability is really focused on reducing the unsustainability of a flawed economic development system that is increasingly based upon an addiction to commodified, material consumption."[38]

Avoiding unnecessary consumption altogether has a greater positive effect on the environment. It may have some short-term negative impacts on economic and social dimensions of sustainability if practiced widely, but the economy will adjust.[39] If we ought to protect the environment, then we should cultivate frugality skills that support reduced material consumption. Many people believe we have obligations to leave enough habitat for other species to flourish. For them, the habitat destruction and biodiversity decline resulting from our consumption habits provide a powerful reason for cultivating frugality skills.

Even those who are skeptical about obligations to protect the environment should be concerned about the unfairness of consumption habits in wealthy nations. Consumption levels in the US could not be exported to developing nations without rapidly speeding environmental degradation and immediately threatening the welfare of a large portion of the human population. The ethics of just consumption deserves more discussion; the interested reader may wish to start with works by Crocker and Linden; Singer, Barnett, and Cafaro; and Newholm.[40] Those who want to look at just basic information on the topic can consult the University of Michigan fact sheet on the per capita US environmental footprint looking at food, energy, water, material use, and greenhouse gases. It estimates that if the current global human population lived like Americans, our environmental footprint would be approximately five earths.[41] There is room for considerable difference of opinion about what global justice requires regarding consumption, but it is hard to find a plausible account of justice that permits the current distribution of consumption.

These moral arguments are commonly understood as based on principles that we should avoid causing unnecessary harms to others. But we know that reducing our individual consumption does very little to reduce harm. From a consequentialist point of view, it is easy to rationalize a bit more consumption because its impact will be negligible. For example, many think they ought to buy only fair trade, shade-grown coffee. But if just a few of us purchase accordingly, our actions will not affect coffee-growing practices. However, we can each serve as a model for our friends, family, and acquaintances (flawed models, to be sure). But if enough others cultivate their own version of frugality virtues, then the impact of frugality magnifies, and our culture becomes more sustainable. While we can each contribute a little to actualizing this goal, we need not depend on its being reached for the moral arguments to serve as powerful motivators.

Given the combined reasons for cultivating frugality and the increased attractiveness of constructive frugality, we might expect that frugality skills would be on the rise, especially among young people who are highly concerned about our approaching planetary boundaries. Yet, despite some pockets of frugality enthusiasm, overall material consumption levels have not retreated. I suspect that the primary barrier to serious cultivation of frugality is the widespread belief that technological advances will enable increasing material consumption for all.

Technological Fantasy?

Technology has substantially increased our ability to live within our planetary boundaries, so why suppose that will not continue into the foreseeable future?

Andrew McAfee argues that technology is currently enabling us to continually reduce our impact on the planet, in effect enabling us to get more goods from fewer materials; he thinks that technology and capitalism are the main drivers of this progress, which he thinks will continue.[42] To justify shifting our emphasis to frugality skills, we must have good reasons for thinking that purely technological solutions for sustainability challenges are likely to be a fantasy. In short compass, my reason is that the power, scale, and speed of technological fixes to global problems are likely to have unintended side effects that destabilize other aspects of the socio-ecological systems on which we depend. As we fix one problem, we create another very soon and on a scale that is equally problematic.

We know that even the most beneficial of our technologies have negative side effects. Indeed, it is hard to imagine implementing any technological solution without incurring risks. Take the example of medical technologies, most of which have strong benefits. With drugs and medical procedures, we are told about potential negative side effects (admittedly often in fine print). These are foreseeable and quantifiable, but many longer-term systemic risks are not easily anticipated or quantified until a problem surfaces. We learn that trace amounts of medications can be found in our post-treatment drinking water, creating other potential health risks, or that the ease of treating mental health issues with pills has led to overprescription and lack of development of alternative treatment approaches.

Still other risks come from the long-term results of the technology achieving its intended goals. As we prolong people's lives, we increase population pressure on the environment, and as birth control becomes more available, we shift toward an aging population, which creates long-term economic risks. Other systemic risks come from the accumulation of power in medical and pharmaceutical companies, whose interests sometimes diverge from those of the populations they serve, as in the case of the OxyContin addiction epidemic. And then there are risks that come from shifting our cultural patterns of interacting with others and the natural world, which potentially impoverishes our lives. Arguably, the medicalization of death has shifted our relation to the dying and reduced the likelihood of having a "good" death surrounded by family. This incomplete sketch of systemic risks of drugs and other medical advances illustrates why even the most beneficial technologies typically create a host of problems that demand further technological development, leading to one version of the technological treadmill.[43] Many technologies are less clearly beneficial but still create a cascade of problematic systemic impacts.

The standard response to this kind of argument is that when negative side effects emerge, the free market will reward those who mitigate the negative effects. The incentives for investing in the innovations that address these side effects are high, so solutions arise fairly quickly. But this typically works only with a subset of problems, those that have a specifiable solution and that arise at a scale, speed, and level of complexity that can be addressed by current institutions. It rarely works with wicked problems, as explained in chapter 1. As our technologies become more powerful and their adoption is rapid and globalized, their negative side effects tend to become wicked problems that can only be managed through negotiated agreements and cultural shifts even when accompanied by further technological development.

The problem of climate change provides a clear case in point. On small scales, coal and natural gas used by some early cultures as energy sources caused few intractable problems. After the Industrial Revolution we needed much more energy, which was easily supplied by fossil fuels that had high energy intensity and were cost-effective to acquire. As the by-products of fossil fuel burning became a pollution problem, the development of pollution-control technologies like scrubbers in smokestacks and catalytic converters on cars helped address these problems (though because of policy changes, not just free-market incentives). This enabled greater use of fossil fuels with fewer immediately obvious negative side effects.

We made massive investments in infrastructure supporting fossil fuel use, such as coal-powered electric plants, our highway system, our aviation system, and a plethora of gas-powered tools. This tended to lock us into a fossil fuel energy system. The levels of twentieth-century CO_2 emissions began to acidify our oceans and to warm the earth. When it became widely appreciated that CO_2 emissions were increasing rapidly, our infrastructure commitments, the power of the oil and coal lobby, and skepticism about governmental regulation made it politically difficult to address the issues. Climate change is a paradigmatic wicked problem. Renewable energy and electric transportation are often touted as the new technological fix, but the speed and scale of implementation have been inadequate to the need to keep average temperature increases below 1.5° Celsius. We would need a great deal of individual behavior change, economic support, and enforceable global agreements to supplement our renewable technology development. The free market will not do the job.[44]

Moreover, massive shifts to renewable energy raise a host of new questions about potential negative side effects. Wind and solar energy are intermittent; we need either energy storage or a supplementary energy to meet demand. Can we store enough renewable energy in battery systems to provide energy when the wind does not blow and the sun does not shine? Will the creation and disposal of huge battery systems lead to other environmental or health issues? Will we need new investments in nuclear energy, which has an intractable nuclear waste problem? Does the increased mining of zinc, copper, lead, and lithium necessary to create these technological fixes actually expand our use of energy, requiring even more renewable system development? Michael Moore's controversial 2020 film, *Planet of the Humans*, explores some of these questions. Many have criticized details in the film, but his point remains that it is unlikely there is a renewable-energy technological fix for climate change unless it is conjoined with massive lifestyle changes. For similar reasons, carbon capture and storage at the requisite scale is likely to create other wicked problems that require culture change, not just more technology.

For climate change and other large-scale global problems, technological changes are certainly important, but they are insufficient. More importantly, the scales at which they must be used create social problems that cannot be solved technologically. The green revolution in agriculture, the use of computers to increase efficiency in the schools, and many other examples confirm this general point. If we assume that new technologies will solve our problems, we are unlikely to undertake the more arduous task of shifting who we are as a people. And as each of us tries to navigate the challenges of our times, we cannot wait for the next technological innovation to figure out how to flourish amid the turbulence. Frugality skills will be highly valuable no matter what technologies emerge.

A Frugality Journey

What would it look like for us to buck the current norms and become frugal activists? My own story provides one possible answer and a few lessons. Early on, necessity played little role in my frugality journey. I was born into relative privilege, in the suburbs of New York City. My parents, who had been children during the Depression, were fairly frugal, but like so many in my generation, I rebelled against them, and especially against their focus on long work hours and deferred gratification. As a teenager, I was expected to find summer jobs like mowing lawns and painting houses, but I strongly preferred to work as little as possible. Walking in the local woods became my favorite form of free entertainment. My small circle of friends reinforced early skills of making do and creating our own fun.

In college, I fell in love with philosophy and found that the pursuit of wisdom required few resources other than some used books. My enjoyment of cognitive goods blossomed. I cared little that career paths for philosophy majors were few and underpaid—no doubt another symptom of my privilege. Much to my father's dismay I bypassed the law school that had accepted me, and instead pursued a PhD. Frugality was a necessity in graduate school. There I began to strengthen the self-restraint skills that would support the long slog toward a tenured position. Finally, I had a strong reason to defer gratification.

As luck would have it, my grandparents had a sheep farm in upstate New York, which I had enjoyed as a child. During graduate school, some friends and I built a cabin there. I reveled in learning how to work with my hands. Local farmers helped us and taught us a great deal, as did my great uncle, who had been a boat builder. We grew much of our food, cooked over an open fire, and explored the landscape. I began to develop the skills of finding beauty in my surroundings and building supportive relationships with neighbors. At first, this was a diversion from graduate school; later it became a way of life. My motivation for cultivating these skills was primarily prudential.

I had been very shy growing up. I felt most at ease in nature and grew to love the outdoors. The environmental movement influenced me in college, but I did not become an environmentalist until later, during my first teaching job in rural North Carolina. There I was surrounded by cotton fields that were defoliated every fall during harvest. A low-level radioactive waste site was slated for development in our county, which we successfully fought against. I started teaching environmental ethics, which was transformative for many of my students and for me. I began to emphasize the moral arguments for reducing material consumption in justifying my frugal behavior and to craft a personal narrative that integrated restrictive-frugality skills with the social, aesthetic, and spiritual skills I had been developing. This narrative played an important role in reinforcing frugality habits when I started to have some discretionary income to spend. Although my starting salary for my first full-time teaching position was only $13,000 ($37,240 in current dollars), the habits honed in graduate school made that more than enough for living comfortably.

The temptations to live in accord with the dominant material abundance norms grew, however, as I advanced in my career and had a child. My extended family embodied these norms, and necessity no longer required that I avoid them. Why deprive myself of consumption pleasures that I can afford? This question has been a constant companion over the last thirty years. My answer has been: because morality requires decreasing environmental impact and promoting consumption justice, and because my frugality skills have become increasingly central to who I am.

Several lessons emerge from the story so far. Typically, mixed motivations, fortuitous coincidences, and indirect pathways contribute to cultivating frugality skillful habits when necessity does not demand them. We often practice relevant skills without focusing on their link to frugality. The same point applies to other clusters of skillful habits we have explored. To weld them into a character trait, however, we also need to develop a personal narrative that synthesizes habits acquired by happenstance into our self-image. This kind of narrative reinforces the process of refining the skills and deepens our motivations for applying them.

In mid-career, I moved back to upstate New York to live on my grandparents' farm and teach at nearby Green Mountain College. I soon became dean and then provost there and had the responsibility to try to envision and enact a sustainability-themed liberal arts education on a shoestring budget. Now frugality emerged as an institution-level priority, but the tradeoffs with investment in people and sustainability innovations became much more salient. Self-restraint no longer answered the key questions, since the puzzles were about investment and allocation strategies. The kinds of judgment involved in advanced frugality skills became crucial. Penny-pinching in the wrong places was as bad as extravagance.

At the same time, I had more discretionary income than I ever expected to have. Almost worse, the time constraints of my administrative roles made my self-provisioning and making-do habits seem questionable. I did spend and consume more, especially in the social networking and traveling that came with the job. Was I slowly abandoning frugality, or adapting it to new circumstances? The puzzles about what frugality requires became highly salient. I began to adjust my judgments about various balances involved in touching the world lightly— for instance, the balances between savoring material consumption and pursuing intangible goods, between saving time and saving money, and between deferring gratification and enjoying the pleasures of the present.

Of course, the tensions persisted, and the danger of self-deception seemed ubiquitous. For example, as I presented a frugality paper in Los Angeles, I wondered aloud whether the expenses of travel to the conference and a nice dinner afterward with friends violated the norms of frugality. If we say that such forms of luxury consumption are compatible with frugality, then it is not clear why buying a small yacht and a personal jet are not frugal purchases for a rich person. The kind of consumerism that frugality was supposed to counteract now seemed potentially compatible with it.

I did go out to dinner with friends after that talk, when I could have bought some locally produced apples and cheese for a light meal. My friends would not have enjoyed a picnic, especially as there was no place to have it. I could afford the meal and would not seek college reimbursement. I weighed these elements and many others, not in some complex calculation, but as a set of improvisations on the theme of touching the world lightly.

The lesson that this part of the story conveys is that judgments about how to express frugality in a context are often complex and debatable, especially where they articulate a balance between appreciating some material good and emphasizing frugality. This balance involves the binocular vision we explored in prior chapters. Such judgments require weighing the impacts of different actions in an imaginative process to see which best embodies touching the world lightly in that context. Steve Fesmire emphasizes this role of imagination when he likens moral decision-making to jazz.[45] The process is highly responsive to the context of decision-making. It is a social activity, involving empathetic skills to envision how others will interact with a riff. It requires creative improvising and adjusting to the reactions. But of course, it is not just impromptu experimentation. It is based in a refined set of habits developed over time that enable one to harmonize the disparate factors in a social context to create an elegant solution for touching the world lightly. Here we do not usually find a single correct view of the demands of frugality, but only better and worse interpretations.

Throughout my time at Green Mountain, I continued to practice most of the self-provisioning skills suited to my region. I cut wood for heat and grew all the vegetables and meat that we ate at home. I invested in solar panels that covered all household electricity needs. Much of my entertainment came from friendships, forays into the surrounding farmland, and occasional camping trips. In semiretirement, I have more time and less income, so the delicate balances have shifted. As noted, I appreciate more beauty on a daily basis, but I am also more inclined to irrationally hoard old shoes and duct-taped work gloves. I spend more time learning how to fix things that I should probably send to an expert. I must be vigilant to refine my judgments about the limits of reasonable frugality. My narrative attempts to integrate the various skillful habits of frugality and make such judgments easier, but it certainly has not eliminated judgment errors. They are inevitable. Skillful habits must evolve as circumstances change.

Our personal narratives reveal how we attempt to resolve the tensions between our various enterprises and the skillful habits that enable them. They highlight questions we struggle with, and they reinforce development of the skills needed to achieve our goals. Of course, they can be flights of fantasy, detached from reality and laden with self-deception. Or they can be well grounded and highly salient stories of who we really are. If our narratives are to reinforce the skillful habits we need, they must be brutally honest and rooted in our best understanding of the systems within which we operate.

We can encourage that honesty by creating practices that regularly remind us to focus on evaluating how well we are living up to our narrative. Ben Franklin tracked in a small book each time he failed to live up to the virtues he was trying to cultivate.[46] At the end of each day, he would survey his activities and note any faults he observed on the page designated for the virtue in question. In this way over time, he could assess his progress in eliminating faults and perfecting his virtues. Sonia Sotomayor always asked herself two questions before going to bed: "What have I learned today? What have I done for others?"[47] In reflecting on her answers, she kept her focus on continuous learning and on helping others. Some people put notes on their refrigerators; others rehearse quotations that evoke a habit. We need practices that stimulate the reflective skills discussed in chapter 3.

To root our narratives in the context in which our lives play out, we need to emphasize the development of our systems-thinking skills, which are the subject of the next chapter.

LEARNING TO THINK LIKE A MOUNTAIN

When the *Exxon Valdez* oil tanker ran aground on Bligh Reef in Prince William Sound in Alaska, it immediately began leaking oil into the sound. According to the oil spill mitigation procedure that had been approved by the Alaska Department of Environmental Conservation, the Alyeska Company was to mobilize oil containment and begin cleanup within six hours of the spill, but the initial response took more than fourteen hours. Confusion about cleanup processes and who was in charge affected the oil containment process over the next two days. After a storm came up four days later, 10.8 million gallons of leaked oil from the *Exxon Valdez* had created an oil slick over nine hundred square miles of ocean and poisoned thirteen hundred miles of coastal shoreline.

That was 1989, but for years afterward the question of who was to blame for this catastrophe received considerable media attention. Many people fixed on the master of the ship, Joseph Hazelwood, who had been drinking before leaving port and who left Gregory Cousins in charge of navigating around some ice while he went belowdecks. Many others blamed Exxon for poor supervision and understaffing. Some blamed Alyeska for failing to contain the spill early on. We could also blame legislators/regulatory agencies who failed to require use of double-hulled tankers, a problem they rectified in 1990. We could also blame citizens who continued to drive gas-guzzling cars and did not support the transition away from fossil fuels in large enough quantities. Perhaps Cousins should bear some blame. The list goes on. Very few stopped to ask whether this kind of blame game is useful in the first place when causal explanations are complex and systemic.

Of course, we should hold people responsible for illegal acts, and it is natural to want both retribution and restitution for harms that are caused by others. But Americans sue others at a much higher rate than in any other developed nation.[1] When something bad happens to us, we tend to seek out an individual, group, or institution who should shoulder blame, and we demand redress. Psychologists have found that numerous factors increase our likelihood of seeking to blame others, including low self-esteem, self-protection, and the influence of other blamers.[2] But at its root, this habit flows from a form of individualism that is embodied in our views about causation and personal merit. We tend to simplify causal stories, looking for discrete actions or events that constitute *the* cause of some result. We are not nearly as good at seeing causation systemically, as a function of the dynamics of many interacting parts.

Understanding causation in terms of systemic dynamics is particularly important in the age of climate change. Major problems like climate change, biodiversity loss, social polarization, income inequality, and racism cannot be effectively addressed without systems thinking, as we saw in chapter 1. My use of the adaptive cycle as an explanatory framework is a piece of systems analysis. Moreover, our ability to flourish now will be enhanced if we can understand and navigate the dynamics that create opportunities for us. The individualistic patterns of thought emphasized in our culture are ill-suited to meeting our challenges. Shortly I will expand on these reasons for strengthening systems-thinking skills and developing a binocular vision that enables us to integrate individualistic explanations and systems analysis.

Thinking Like a Mountain

Most environmentalists are familiar with Aldo Leopold's evocative metaphor "thinking like a mountain."[3] Leopold was concerned that when we focus just on the relations between a few parts of an ecosystem, we often fail to understand the complex interactions that make the system work. In an iconic story, he describes learning this lesson after trying to increase deer herds by killing off wolves. The deer herds did grow, but as a result they exceeded the carrying capacity of the land, browsing all edible vegetation until they died of starvation. Thinking like a mountain involves understanding the land as a community or system, whose members play important roles in its functioning.

To understand systems-thinking skills, it helps to have some theoretical vocabulary; in this respect these skills differ from the others we have explored. The theory of complex adaptive systems provides a general description of how self-organizing systems like ecosystems, organizations, and political bodies function. Such systems "involve many components that adapt or learn as they interact."[4] They have emergent properties that cannot be understood in terms of properties of their components. Linear causal relations between their parts cannot fully explain their behavior; complex feedback loops play an important role in systems functioning. Many complex adaptive systems are self-organizing, which means that they function in ways that preserve and often enhance their own functioning rather than succumbing to increased entropy over time.

Such systems concepts are common in some disciplines, but the associated skills for thinking about systems are rarely taught in any depth. My discussion will focus on the skills, but only by providing enough detail that a reader can understand the reasons for thinking that we all need to strengthen these skills in order to flourish now. Several excellent overviews of systems thinking are already available.[5] Small-scale systems like families, work environments, and local ecosystems are where most people can learn to apply systems thinking to make change, so I will illustrate the skills with such examples. The next five sections identify specific systems skills.

System Description Skills

How do we circumscribe a system that we want to understand? This question is often devilishly difficult to answer. Many systems are highly permeable, with energy and objects moving in and out of them. Donella Meadows, a renowned systems expert, says, "There are no separate systems. The world is a continuum. Where to draw a boundary around a system depends on the purpose of the discussion."[6] To be good at systems thinking, we need to be able to describe systems' spatial boundaries in ways that are useful in understanding their dynamics and leveraging change.

Of course, some systems, such as a city or a lake, seem to have clear geographical boundaries. For practical purposes we may use these boundaries, but they can be misleading. For example, it is tempting to describe a college system in terms of its property boundaries, but many of its activities extend far beyond these boundaries to include suppliers, the commutes of workers, projects done by students in the local community, and the work of its online students around the world. Trying to understand a college as a system involves making judgments about what activities to include for what purposes.

Imagine a food co-op called Mandy's, which has a storefront in a small town, five part-time employees, and one full-time manager. The nonprofit co-op is struggling to survive, and the board is trying to analyze the system to leverage positive change. Initially, stakeholders might begin thinking of the system as including the employees, the customers, the products the co-op carries, and its board of directors. A consultant suggests that they also include in the co-op system the suppliers of products and other townspeople who do not yet shop at the co-op. The latter might be controversial, but if the system description does not include those who might be induced to shop there, that may affect the range of solutions to the co-op's low revenue problems. We must add prices of products, since that affects sales, as do the location of the storefront, the percentage of people in the community who can afford to shop there, the competitors in the region, and numerous other potential system components.

Finding a good balance between simplification and complexity is another skill of system description. Dialogue between stakeholders can create a more complex view of the system. But if the picture gets too complex, it will be hard to understand its core dynamics. As Walker and Salt note, we must always ask, "What is the minimum but sufficient information we need to incorporate in our understanding—our models—to make robust decisions about planning and management?"[7] The board at Mandy's needs to be careful not to try to include everything about the co-op in its model.

The skills of systems description also include identifying the functional roles of system components and the relationships between these parts. In simple systems, the parts may be obvious, like the key stakeholders in a town, but in others there is an art of identifying which parts have crucial causal roles in system functioning. Formal systems modelers often identify key stocks of goods (e.g., water in a reservoir or trust in a community) and flows that affect the levels of these stocks. This enables us to see how different parts of the system affect other parts. Later in this chapter we will see how we can use stocks and flows to refine our understanding of the sustainability of a community.

When we are parts of the systems we want to understand, we must pay special attention to our influence on system behavior, including our unintended contribution to its problematic dynamics. David Stroh highlights the skill of analyzing our inadvertent contribution to systems problems we are trying to solve. As he notes, "If you do not understand your role in the problem, it is difficult to be part of the solution."[8] Take the system of mass incarceration, for example. In trying to diminish crime in America in the 1990s, many people supported more aggressive policing and longer jail terms, which eventually led to people being released from jail who had few options other than returning to crime and thus exacerbated the problem. Those supporters inadvertently contributed to the problem of crime and the many other negative side effects of mass incarceration.

Feedback Analysis Skills

As we refine the above skills of system description, we need to build our capacity to identify the feedback loops that govern system functioning. Kastens and colleagues put the point succinctly: "Because feedback loops underpin a stable Earth system, fostering a working knowledge of this concept throughout the decision-making populace could increase civilization's capacity to cope with 21st-century challenges."[9]

Some feedbacks help to maintain an equilibrium in the system. When a system moves outside of an equilibrium state, then a sensor triggers a sequence of events that returns the system toward equilibrium. A thermostat or a car's cruise control exhibits this kind of "balancing" feedback loop. Often the results of balancing feedback are desirable; our body sweats when overheated to reduce heat through evaporation and to return the body to its equilibrium temperature. However, sometimes balancing feedbacks preserve a system in an undesirable state. In Mandy's co-op, the difficultly of keeping a broad inventory of products when revenue is low may cause people to shop elsewhere, which keeps the system in a low-revenue state. The board must find a mechanism for leveraging change in the system that will move it out of this negative equilibrium. Ecosystems overwhelmed by an invasive species face similar challenges. We need to break the feedback to leverage change in the system.

Other "reinforcing" feedback loops are important for system growth. Here the feedback accelerates system change in either a positive or negative direction when the output of a process feeds back into the process, further amplifying output. In other words, reinforcing feedbacks create either virtuous or vicious cycles. In the case of Mandy's co-op, the board is trying to find an intervention that will not only build short-term growth in revenue but will start a cycle in which each new growth stimulates further growth. Jim Collins calls this "the flywheel," which is difficult to get moving, but once it is moving with some speed, it continues to accelerate because of its own momentum.[10] He argues that businesses that create and patiently nurture a flywheel can become great. Apple might be a current example, with its increasingly sophisticated versions of the iPhone. Vicious reinforcing feedbacks are particularly troubling because they can lead to system collapse if they are not stopped. Climate change involves numerous vicious cycles. For example, as permafrost melts because of increased air temperatures, it releases methane, a potent greenhouse gas, which further increases temperatures, releasing more methane, and so on.

Often feedbacks have delays between the input and the output of a feedback loop. For example, if Mandy's starts advertising sales for products to increase revenue, it may take months before we can tell whether the input has the desired effect. The delays that characterize our climate system, as well as its scale and the huge number of actors, make it hard for us to see the effects of our behavior. One way of intervening in a system is to try to shorten delays in feedbacks that govern the information we receive from behavioral inputs. This can shift behavior much more reliably than delayed information.

We need to become much better at determining where reinforcing and balancing feedback loops are influencing system behaviors and at learning how to shift feedbacks. Simple feedbacks are often easy to see, but in complex systems, feedbacks may interact in ways that mask their characteristic behaviors and make them hard to identify. System modelers create diagrams that highlight causal relations between key system components, including feedbacks; see figure 4 for a simple model illustrating the dynamics of the homeless population in response to various interventions. We do not necessarily need to learn how to create such diagrams, but we do need to learn how to see feedbacks in the systems that affect us.

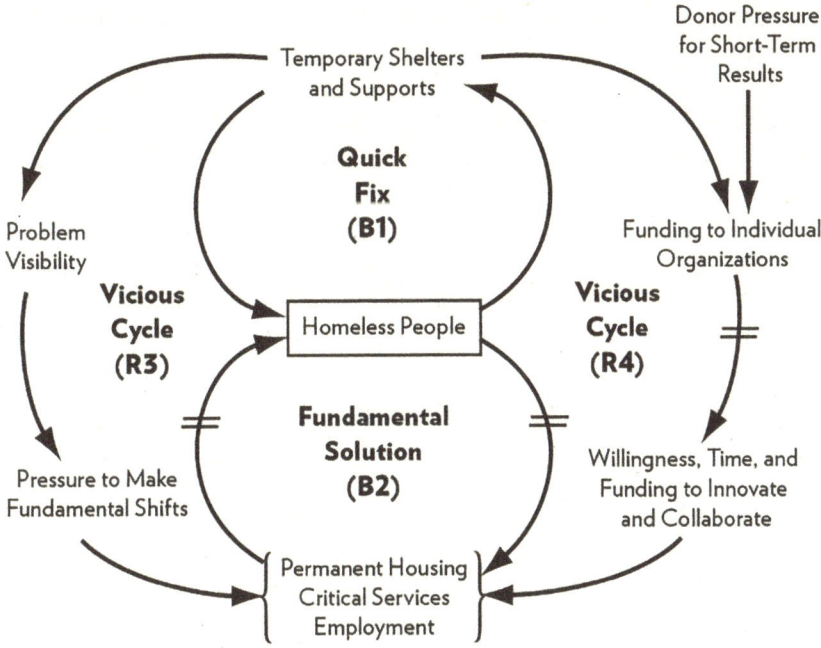

FIGURE 4. A systems diagram shows feedback loops between homelessness, temporary shelters, and permanent housing. Reprinted from *Systems Thinking for Social Change,* copyright 2015 by David Peter Stroh, used with permission from Chelsea Green Publishing (www.chelseagreen.com).

Threshold Analysis Skills

As we saw in chapter 1, when a system crosses a major threshold, its structure and functions change. We commonly call thresholds "tipping points" because the system tips into a new trajectory. Sometimes the resulting changes fundamentally shift the identity of the system and are irreversible. In other cases, these changes alter the behavior of a system and move it into a different regime.[11] A lake that receives too much phosphorus-laden runoff can tip into a eutrophic state with frequent algal blooms, but it can also tip back with remediation.

The 2004 disaster film *The Day after Tomorrow* dramatizes what could happen if the Atlantic Meridional Overturning Circulation (AMOC) system, which governs the Gulf Stream, crosses a threshold. If climate change causes enough of the Greenland ice sheet to melt, the resulting influx of cold fresh water could shut down the Gulf Stream, causing temperatures along the eastern US coastline and northern Europe to plummet. Researchers do not know exactly where the thresholds lie for the AMOC system, but recent studies suggest that we are eroding buffers that prevent us from crossing such thresholds.[12] Only an expert could be expected to estimate key thresholds for AMOC, but all of us should increase our capacity to estimate the location and kind of thresholds that might govern common local systems.

Economic and ecological thresholds are usually more easily estimated than social thresholds (e.g., low trust or high conflict thresholds in organizations). In the Mandy's co-op example, we know that if revenue drops to the point that employees cannot be paid or loan payments made, and no additional funding can be secured, the co-op will need to declare bankruptcy. A seasoned manager may be able to shift such a threshold a bit by cutting other expenses, but that manager will know generally how close the threshold is. We can learn to recognize changes in the dynamics of a system as it nears a threshold. We can also learn how to assess the amount of reserves or buffers a system has that enables it to avoid crossing thresholds. Such skills grow with increasing familiarity with the kind of system in question, and they are partially transferable to similar systems.

I often teach systems theory skills to undergraduates using close personal relationship examples. Students have a general understanding that relationships can cross a threshold beyond which the relationship collapses or changes dramatically. Conversational feedback loops, shared positive experiences, and couples counseling can keep a struggling relationship system alive. However, if one or both members of the relationship arrive at a stable belief that the benefits of keeping the system going are not worth the effort, the relationship crosses a threshold and will dissolve. We can generally describe what this looks like, but in a specific relationship we may not know when such a threshold will be reached.

Understanding roughly where a threshold is also helps us assess when to nudge an undesirable system over a threshold and into release. If a system needs significant transformation to meet the needs of key stakeholders, then it becomes imperative to find interventions that can tip the system over a threshold so that significant reorganization is possible. This is often dangerous, since one cannot know whether a desired transformation can be achieved. Release can create enough chaos that the situation gets worse rather than better even in the medium term. Yet we can also learn the skills of navigating release and diminishing the severity of its impacts.

Leverage Point Assessment

The big payoff for learning the above skills is that we become able to leverage change in a system. In a complex system, we usually do not have enough control to simply create the changes we want. Even a powerful leader of an organization cannot wave a wand and make the organization function in a more productive manner. The leader needs to be able to identify places where small changes can be made that will lead to a large desirable change in system functioning. In developing this skill, we must learn how to distinguish between strong leverage points and weak ones. Often well-intentioned interventions have short-term gains but counterproductive consequences in the long term (what Senge calls "a fix that backfires"). In Mandy's co-op, expanding the number of products in stock can have a quick positive impact, but it can create cash-flow and quality-control issues if the store does not have the capacity to manage the full set of products, which may ultimately turn off customers.

Donella Meadows's *Thinking in Systems* contains the classic discussion of weak and strong leverage points.[13] She notes that "leverage points frequently are not intuitive. Or if they are, we intuitively use them backward, systematically worsening whatever problems we are trying to solve."[14] She provides a list of twelve kinds of leverage points, ordered from weaker but easier to change to stronger levers, which tend to be harder to change.

Parameters such as the rates at which something occurs are fairly easy to shift. We often focus on these to address a problem, for example shifting the sentencing guidelines for crimes, or adjusting tax rates or subsidies. But Meadows argues that these are unlikely to fundamentally shift the functioning of a system, so they are typically weak levers. Changing the rules governing a system or its goals is more difficult; it requires more power and concerted effort over time. The results of such shifts tend to significantly alter feedbacks and other aspects of system functioning. The skills associated with finding the best levers for shifting an undesirable system are linked to threshold identification skills.

For Mandy's co-op, the problems seem large enough that the board may need to consider a fundamental change in goals to shift the way the co-op functions. Perhaps adding a small coffee shop in the storefront (if the town lacks such a gathering spot) might drive an increase in customers, who would also be more likely to buy goods that the store regularly keeps in stock. This plan would require reallocating staff time and floor space, but it might enable the co-op to stay afloat and keep many of its services to the community.

In close personal relationships, talking about problems is by itself a weak leverage point. It is often done poorly, it rarely addresses underlying issues, and often creates resentment. But conversations that shift rules governing behavior in ways that enhance the experience of both partners can breathe new life into a relationship. Finding such solutions may be difficult and take practice to implement even when they are initiated. Effective leverage points often take time to work, so making small wins salient early in the process is important.

My formal education did not teach me anything about systems skills. It was only in mid-career that I began to learn how to effectively navigate and alter dysfunctional systems. I had to find places where I could experiment with fledgling systems-thinking skills. The courses I was teaching provided a great opportunity.

I identified courses as systems that lasted for a semester, had a clear part-time membership, and developed a dynamic pattern that often determined the success of the course. I saw how balancing feedbacks tended to constrain efforts to shift problematic dynamics and how negative reinforcing feedbacks could lead to a course crossing a threshold beyond which most students tuned out and just did the minimum to get by. I used systems concepts to explain a problematic dynamic to students. For example, I called out free-rider issues, where some students were doing the work and dominating the discussion while others were sitting in the back row and taking shortcuts to try to get a reasonable grade. As the class grappled with how to minimize free riders, I noticed that I was receiving much more sophisticated suggestions about changes we could all make. I thought of teaching a course as leveraging some change in a system, not just providing an educational experience. The gold standard became creating a positive reinforcing feedback loop during which students kept trying harder and harder to learn from each other and from the instructor because every increase in their time investment provided more learning that they found useful. When that happened, the course seemed both important and fun.

Common-Good Skills

Systems-thinking skills naturally support capacities for promoting the good for whole systems. Aldo Leopold paired his views about thinking like a mountain with an ethical view that we ought to preserve the good of whole ecosystems, which he called the land ethic. His shorthand principle for this ethics is that we should preserve the stability, integrity, and beauty of the biotic community.[15] Similarly, some ethical theories recognize a duty to support the common good, understood as more than the sum of individual goods. Common goods in human communities usually include their education systems, road systems, parks and recreation areas, systems for administering justice (e.g., courts), clean air and water, and so on. Theories about common goods are often built on views that understand communities as whole systems that should be valued as such. But what skills do such holistic ethics underwrite? I suggest we break these into two groups: skills involved in evaluating potential common goods, and those used in promoting common goods.

When we are not being careful, we may assume that what we want for a community or organization is part of the common good, but that is often not the case. It is easy to confuse our desires and the common good. To identify common goods, we need the skills of detaching from our desires and determining what would benefit most people in the community. This involves learning about the whole community (not just subgroups), assessing the feasibility of different potential common goods considering community constraints, projecting the impacts of larger-scale systems on the community, and determining what tradeoffs make sense in the context.

Community planning provides an excellent practice field for such skills. As I write, I serve on a committee in charge of developing a ten-year plan for my rural town, Hebron, New York. We have consultants who supply us with data and planning expertise, and we have created a fairly robust public input process, but the committee has a lot of freedom in deciding what goals the plan should prioritize, what would best serve the common good in the town. I started the process thinking that the town should develop some recreational trails that would highlight the beauty of the region, enhance economic development, attract more young people, and improve public health. I am an avid hiker, so I see the benefits of trail systems in other towns, and I would use the trails regularly. But during our process I learned much more about the interests of residents, the difficulty of acquiring trail easements, and other town needs that had priority. I realized that even though a trail system can be an important common good, it would probably not be an important one to pursue for my community.

Promoting common goods involves marshaling support for aspects of a community that are regularly taken for granted, diplomatically calling out processes that are eroding the commons, creatively integrating pursuit of personal and common goods, and setting an example of how to give back to a community through volunteering and leadership. With only a handful of paid employees,

Hebron will be successful in implementing its plan only if community members step up to do the work and have the skills to do so effectively. But will that happen?

We are inevitably torn between pursuing personal goals and community goods. So many people in small towns are overwhelmed by the demands of meeting their own basic needs, caring for family, advancing at work, and helping neighbors that they have little left to give for the common good. Many others just do not care. And yet, unless common goods are recognized and prioritized by a critical mass of people in a community, the system as a whole will tend to erode and become less capable of providing the goods that residents depend on.

We each need to find a balance between pursuing our own personal goods and promoting common goods; ideally we find creative ways to integrate some of these aims. Unfortunately, our cultural emphasis on individualism and pursuit of personal advancement makes it hard to find that balance. We vote our views, pay our taxes, and occasionally pay attention to how well our communities are functioning, but this is far from sufficient in the age of climate change, where many common goods have been severely degraded. In larger-scale systems, many people will not be able to do much more than vote and pay taxes, but that is why we must focus on smaller-scale systems—our neighborhoods, our workplaces, our churches or clubs—where we can join with others to alter systems in significant ways. These local changes can then be scaled up if the conditions are ripe.

Systems thinking involves a host of other skills, including humility and collaboration. A good systems thinker easily adopts multiple perspectives on the systems in question and recognizes where blind spots about system functioning may occur. This requires the humility skills outlined in chapter 3. Often the collaboration of different stakeholders reveals system processes that an individual alone would miss. Thus, the collaboration skills described in chapter 2 are also important. Mastering systems thinking is a lifelong endeavor. We need not become experts in order to improve our chances of flourishing, but we must become much better at these skills and make their use habitual.

Integrating Systems Skills with Individualism

In prior chapters, we saw that strong cultural norms reinforcing clusters of skillful habits often make it hard to acquire contrasting skill sets. Our individualist norms have the same impact on systems skills. The tensions between pursuit of individual self-interest and common goods are a regular part of political debate, as they should be. The tensions between systems thinking and methodological individualism are less salient but still quite powerful.

According to methodological individualism, group behavior is reducible to the behavior of individuals, and we can understand the former in terms of the latter. The skills reinforced by methodological individualism and other forms of reductionism in biology and chemistry include isolating the effects of key variables and testing for correlations with other variables. On this paradigm, if we understand lawlike relations between parts of a system, then we should be able to predict the behavior of the whole system. The contrasting holist view holds that at least some group phenomena can only be understood at the level of the whole system.

The skills of reductive explanation have enabled us to understand and manipulate many physical systems reliably, hence this approach has become almost definitive of science. Its success has been more limited in social systems and ecosystems, where holism has a foothold.[16] While most non-scientists do not think much about reductionism, they are still influenced by the idea that we should be able to find *the* cause of some event. Our individualistic cultural emphasis strengthens a tendency to look for simple causes of complex events, which is manifested in the desire to find someone to blame for complex failures like the *Exxon Valdez* disaster. It also encourages us to address our problems by trying to manipulate salient parts of a system that are taken to be primarily responsible for the problem, instead of finding ways to leverage change in the dynamics of the whole system.

We need norms that positively reinforce both reductionist skills *and* systems skills. We need to develop the binocular vision that enables us to integrate the use of these skills and to use both explanatory paradigms productively. This gives us a richer, more nuanced picture of the world and a larger toolbox for addressing our challenges. In allocating blame for the *Exxon Valdez* oil spill, we may apply our systems-thinking skills to try to understand how the interactions of a large number of factors contributed to the failure. Doing so can help all of us support ways of leveraging change in the system to reduce such disasters. While individualistic skills used to find fault and determine just penalties satisfy a demand for a quick, comprehensible moral response, the systems skills temper such judgments. Binocular vision should also enable the careful crafting of associations that effectively integrate pursuit of self-interest and the common good. As Tocqueville observed almost two hundred years ago, Americans were once very good at building associations that served the interests of the community and the self.

While people with quite diverse viewpoints should acknowledge the importance of this kind of binocular vision, they will differ about where to set the balance point between the contrasting skill sets. Since systems skills are generally underdeveloped, we certainly need to shift the balance point toward them. In the next section, we will see strong reasons for thinking that a habitual use of systems skills will enhance our chances of flourishing—another reason to think the balance needs to be tilted in their direction.

Systems Thinking and Flourishing

Systems skills seem so abstract that it may be helpful to begin this section with a story of how they enhanced my own flourishing. I grew up as a dedicated individualist, inclined both toward pursuing self-interest and toward linear, reductive thinking. Like many in my generation, I flirted with communal living, but I never really experienced a strong sense of community until my first teaching job at a small liberal arts college, where the common good was a central theme. I fell in love with being part of a small, manageable, shared enterprise—a human-scaled community, and I learned how the college functioned.

The opportunities that came my way at a small place developed my leadership skills and also my understanding of higher education systems. Soon after tenure, I chaired the college honors program and then the general education program. The college struggled; we had to make difficult decisions. All of the faculty saw the budgets, and the consequences of poor decisions. We knew viscerally the importance of student recruitment and retention. Without knowing any of the terminology of systems thinking, I began to see how some departments had created positive feedback loops that generated high enrollments, while others languished. I experimented with ways to build such feedbacks in the philosophy department, with mixed success. While the mistakes were unpleasant, I learned as much from them as from the successes.

Fast forward to mid-career after my move to Green Mountain College. Here I focused on teaching in environmental studies and did some program administration. Such leadership roles did little to erode my love of the individual freedoms associated with being a faculty member, but they did shift my balance between individualist and systems skills toward the latter. With great reluctance, when no other options seemed feasible, I agreed to become provost (for a year or two) during a difficult presidential transition. Twelve years later, after serving three presidents, I joyfully returned to the faculty, but I knew I would miss much of a job I had come to enjoy. Having that level of responsibility for the welfare of a community significantly strengthened my systems skills.

I saw my job as facilitating and strengthening achievement of common goods at the school, including the education of all students and the growth of the school. Achieving these goals called on a wide range of social skills to help individuals overcome competing self-interests in order to work effectively together. We needed to craft a high-functioning community with few resources other than our students' tuition. Systems skills provided the framework and analytical tools that helped me create strategic interventions and communicate why we needed to change. Here, analyzing how the college's performance was affected by systems at larger and smaller scales was extremely helpful both in planning and communication. Larger-scale forces like economic downturns, demographic changes, and accreditors' mandates had to be anticipated and effectively navigated. Dysfunction among subsystems had to be mitigated. Shifting problematic feedback loops and developing feedbacks that reinforced effective problem-solving now became existentially important. In the end, it seemed that my most important role was motivating the development of beneficial cultural norms.

But even if I enjoyed the trajectory of my growing repertoire of skills, did it really enhance my flourishing? Despite the increasing stress of more responsibility, the longer work hours, the inevitable failures, I am confident that it did. Honing my systems skills made me happier in my position and better at it. It also made me feel like a better human being. I felt a greater sense of contribution to the common good and more appreciation for what that involves. I felt more competent and confident about navigating the challenges we faced even as I become more sensitive to manifold uncertainties. I was able to adjust my expectations in light of understanding larger-scale system impacts on the college, which reduced disappointments. I became more nuanced in my assessments, less angry, and more caring. I can honestly say that the imperfect and still changing binocular vision I have found has enhanced each of the elements of flourishing described in the introduction, especially positive relationships, meaning, and a sense that my skills were reasonably aligned with the challenges I faced.

The college grew by 40 percent over the first six years I was provost, then it plateaued for a few years before beginning to shrink. It also attained a national reputation for sustainability education, appearing in top ten colleges for sustainability lists from *Sierra* magazine and the Association for the Advancement of Sustainability in Higher Education—surprising achievements for a small rural experimental college. Alas, five years after I stepped down as provost, the college closed its doors, unable to attract enough students or raise enough money to continue operations. I experienced at Green Mountain a large part of the adaptive cycle, writ small—a heady growth phase, a short difficult conservation phase, and then the pain of release. Through it all, the systems skills tempered any tendencies toward hubris and increased my resilience. One silver lining of the college's decline is that I am better equipped now to navigate the turbulence of our times.

The details of this story are mine, but the lessons it contains about practicing systems skills and flourishing are transferable. First, we need to practice these skills on a regular basis in a setting where we can make mistakes and learn from them. We can focus on something we do regularly—for me teaching a course or playing a leadership role—and see how systems skills can help us to do it better and enjoy it more. Second, we should accept some responsibility for leadership even if we do not feel prepared for it. Having some responsibility for how part of a system functions is one of the best ways to deepen our understanding of system dynamics and to experiment with leveraging change. It also deepens the meaning of what we do and at least sometimes provides a sense of achievement. And third, we should work toward being a part of a community whose culture supports a balance between individualist and systems skills. I tell graduates looking for a job or a graduate school to look at the culture of an organization. Will it help to reinforce the skills you want to improve? I was fortunate in this respect.

A story can be illustrative, but we also need general reasons for thinking that systems skills will strengthen our flourishing in the age of climate change. The first of these reasons we have already anticipated; our position on the adaptive cycle calls for greater use of systems skills. The threats associated with planetary boundaries are best characterized in systems language, as are many regional issues like systemic racism and income inequality. Our individualistic cognitive habits make it very difficult for us to intuitively grasp complex socio-ecological systems. As a result, we tend to delay action on our challenges, both because of uncertainty and because our individualistic assumptions often conflict with promising solutions. Such wicked problems have a social dimension, so they are rarely amenable to reductionist, linear problem-solving. To leverage change in most socially complex problems we need to understand the system dynamics that lead to the problem, the likely impacts of interventions, including their potential unintended consequences, and ways that the system we wish to alter interreacts with systems at larger and smaller scales. Solutions to such problems require collaboration among stakeholders, integration of multiple kinds of expertise, and a high tolerance for uncertainty in decision-making.

Furthermore, since we are likely to experience serious release on multiple scales, we need the skills that enable us to successfully navigate the back loop of the adaptive cycle. We have seen that when approaching a release, people experience heightened uncertainty and conflict. Unless release is a result of a large-scale disaster, like a catastrophic flood or a devastating military attack, people will be uncertain about whether the system identity can be preserved. Sometimes release involves a long, unpleasant descent before the system becomes nonviable, which makes it very hard to know whether to hold out for help, prepare for major change, or just to leave. Anyone who has been part of an organization in decline has felt such uncertainty.

Once in release, people will be unsure of whom to trust once established leaders have failed to guide the ship to port. Systems skills help to clarify what is happening and highlight productive options for action. We can expect a great deal more turbulence in our lives as we try to skirt planetary boundaries and navigate local releases. In this context we must prepare for many "mini-releases"—shifts in careers, geographical context, close relationships, and friend groups. Those who have prepared for release physically and psychologically will be better able to face the uncertainty it occasions and to adapt. They will also be able to make better personal decisions about what changes in their own lives may be warranted. They will have a stronger sense of agency.

A second line of argument establishes that we have swung too far toward individualism to flourish culturally and that we need to counterbalance that with a reemphasis on systems thinking, especially common-good skills. Robert Putnam and Shaylyn Romney Garrett's excellent book, *Upswing*, details this reasoning by tracing the shifting social balances we have tried to strike between individualism and community in the United States.[17] They begin with the question of why problems like polarization, economic inequality, and social isolation seem to be consistently getting worse. Their answer is that since the 1960s we have gradually become more and more individualistic, emphasizing personal freedom, individual rights, and self-expression at the expense of pulling together to work toward the common good. The book is an extended, data-rich defense of the thesis that from the 1880s to the present, the US culture is characterized by an "I—We—I" trajectory.

Their historical narrative begins with the Gilded Age of the 1890s, when severe economic inequality, horrendous working conditions, and open class warfare were defended by the individualism fostered by social Darwinism. In reaction, the Progressive movement arose, which emphasized working together for the common good, transforming social norms, and building effective associations. Thus began a growing emphasis on community, which supported the New Deal, massive mobilization during World Wars I and II, and major infrastructure improvements. It also led eventually to the oppressive conformity that many young people rebelled against in the 1950s and '60s. Since then, they argue, individualism has again been on the rise until the present day, as evidenced by survey results, economic data, and a wealth of other data sources.[18]

Although our current form of individualism has enhanced many freedoms, its costs have been considerable in terms of social stability and shared problem-solving. The US response to the COVID-19 pandemic can be fruitfully viewed as reflecting an overemphasis on individualism. Stemming the spread of the virus was a classical collective-action problem. Individuals could contribute to the goal, but addressing the problem successfully required unified action. Since individuals capable of spreading the disease were often asymptomatic, there needed to be widespread testing to reduce the spread. But initially the federal government resisted setting up a testing program, on the grounds that this was the responsibility of states, which at the time were not well positioned to marshal the resources necessary for widespread testing. In various other areas crucial to life, such as clean water and clean air, and defense against enemies, it is the nation state that assumes responsibility for promoting common goods.

At the beginning of the pandemic, leaders emphasized the technical problem of rapidly creating a vaccine for the disease. We excel at solving that kind of problem and succeeded admirably. But the pandemic involved complex social dynamics. Variations in our trust in government, our tolerance for risk, our willingness to shift behaviors and to be vaccinated, made it very difficult to manage the pandemic, especially given the early uncertainty about what kind of management was best and the fragmentation of the public information system. Our emphasis on personal freedom compromised government attempts to create widespread norms that would mitigate the pandemic, leading some to defend significant constraints of freedom (e.g., lockdowns) to promote the common good of public health. This resulted in backlash and significant politicization of pandemic responses. These dynamics partially explain why the US had significantly more deaths per capita from the pandemic than other highly developed nations. We need to be better at addressing collective-action problems at larger scales, and systems skills will help with that.

Putnam and Garrett conclude that once we see the shifting balance between individualism and the common good, we can learn how to become more community-focused by acting on lessons from the Progressive Era. They suggest we need large-scale political change, which requires building coalitions and emphasizes incremental progress over revolution. We need to avoid overcorrection, and we should seek to find ways in which self-interest and common good align.

A third reason for cultivating systems skills highlights the importance of having different kinds of tools in our problem-solving toolbox. Abraham Kaplan coined what he called the "law of the instrument," which states that scientists tend to frame problems in ways that fit the methods or instruments that they have ready to hand. If reductionist methodologies are our primary approach to scientific explanation, then we are likely to use them even where they are less apt, as in addressing wicked problems. This approach creates a serious cognitive bias if we have a narrow set of methodologies. Abraham Maslow and many others popularized this "law" with the adage "If all you have is a hammer, then everything looks like a nail." While we are limited in how many methodological skills we can effectively marshal, it seems wise to supplement our reductionist skills set with systems-thinking skills that approach such problems in fundamentally different ways. This applies not just to scientists but to the culture at large, which must support alternative ways of looking at complex problems. If a culture has diverse toolboxes, it is likelier to be more flexible and creative in its problem framing, which we sorely need in the age of climate change. This applies to individuals as well as cultures.

Lastly, systems skills strongly reinforce the other resilience skills clusters we need to flourish now. We have seen that the complexity of the systems we are trying to shift requires strong collaborative skills and diverse perspectives. This complexity also plays a key role in justifying greater use of humility skills, increasing our open-mindedness about solutions, and decreasing our development of hard convictions that make compromise impossible. Use of systems skills enables us to construct a more vivid picture of our situation on the edge of release that reinforces development of frugality skills to mitigate the unjust impact of overconsumption on other humans and the biota. Systems thinking is the linchpin for navigating our times. These skills enable us to build an empirically grounded understanding of the dynamics of the age of climate change. This serves as a foundation on which the cultivation of other underappreciated skills can be more easily motivated.

Sustainability, Resilience Skills, and Widespread Flourishing

Up to this point my argument has focused on how individuals can maximize their chances of flourishing no matter what happens. Let's turn now toward the other half of the argument—the reasons for thinking that if enough people cultivate resilience skillful habits, then our chances of achieving sustainability increase dramatically, which promises to spread flourishing much more broadly. With systems skills in view, we can now look more closely at what sustainability is and how it connects to flourishing.

By one dated count, over three hundred concepts of sustainability have been advanced.[19] Some experts despair over this proliferation and suggest that sustainability has come to include everything good and therefore means nothing specific. Thus, some advocates argue that we need more precise definitions of sustainability in order for the goal to provide a clear guide for action.[20] Discussions often begin with the Brundtland Commission's definition of sustainable development—"meeting the needs of present generations without compromising the ability of future generations to meet their needs"—but they quickly move to greater precision. To operationalize this definition, we need to know what kinds of needs are at issue, what counts as compromising others' needs, and what tradeoffs between needs are permissible. The United Nations Sustainable Development Goals (2015) codify the range-of-goals problems that sustainability attempts to address, but they do not settle debates about what sustainability is.

The concept of sustainability has both a descriptive component and an evaluative component; it is a *thick* evaluative concept similar to "organic," "natural," "cowardly," and "selfish."[21] Such concepts are used to pick out some attribute in the world that can be empirically identified, *and* they carry a positive or negative value. Both the descriptive and evaluative parts of the concept are essential to its meaning. So, for example, if I say that someone is selfish, I am describing a character trait that is expressed by frequent pursuit of self-interest *and* I am implying that the trait is morally bad. I would use a different word if I did not think the self-interest was bad. Similarly, "sustainable" usually implies both that some elements of the system can be continued over a period of time and that this is good. In contrast to a thick evaluative concept, a *thin* evaluative concept, like "good" or "right," has very little descriptive content; it is almost all evaluative. A purely descriptive concept, like being six meters long, has no evaluative content.

While thick concepts are very useful, they can also create significant confusion. People may align their descriptive and normative content in different ways; what one person thinks of as selfish, another person may think is just rational. They may agree on the facts about someone's self-interest but disagree about whether it is bad or good. Similar things happen with sustainability. Some debates about sustainability are about empirical questions that can be answered by the sciences. Can some activity be continued into the future, given our resources? Or they may be about values. People may disagree about whether it is good or bad that the activity cannot be continued, even though they agree about all the facts. These sources of ambiguity limit the precision of sustainability, but they also make it well suited to characterizing goals that will motivate people.

We can minimize the ambiguity by specifying in general terms what makes a community sustainable. As is common, I understand "sustainability" to address social and economic issues as well as environmental issues. It is not just environmentalism in new clothes. If we use the systems concepts of social, natural, and economic "stocks of capital," we can say that actions reducing a stock of capital below a level that enables a community to retain its functioning and regenerate its stocks are unsustainable. Those that maintain or increase stocks and aid a community's functioning over a timeframe are sustainable for that timeframe.

A couple of examples can help to clarify this abstract idea. The community of Poultney, Vermont, depends on community wells for its drinking water. Any practice of community members that degrades the water coming from those wells in ways that prevent them from supplying clean water to residents compromises the community's sustainability. Likewise, there is an average degree of trust in the town's governance institutions among community members. Practices that degrade that trust too much and make the town ungovernable compromise its sustainability.

This simple model must be modified in at least three ways if it is to capture well the core idea of sustainability. First, sometimes one can reduce one stock in ways that affect functioning in order to increase another stock. For example, the town can take on debt to upgrade the wells so they provide adequate water. The debt may increase risks to financial sustainability over a certain period but in the long run strengthen overall sustainability. Sustainability depends in part on how the stocks of capital interact to sustain the functioning of the whole community. Second, a community's functioning may degrade stocks of other communities while preserving its own, which we will want to say is unsustainable. If Poultney were to dump its sewage into the nearby river, downstream water quality (natural capital) would be degraded. So, we must add that if a community's practices degrade the stocks of other communities in ways that diminish their functioning, then those practices are unsustainable.

The third modification pertains to the impacts that a community's practices would have if other communities acted similarly. Sometimes a community's practices do little to degrade the stocks of capital available for other communities, but that is because others are not acting in the same way. For example, a community might emit pollutants into the atmosphere (like CO_2) that by themselves do not diminish the functioning of the other communities not emitting such pollutants in the same quantity. But if all communities emitted the same amount of pollutants, then the atmosphere of all would be seriously degraded. In effect, we cannot universalize the behavior of the emitting community. This is why even in the early twentieth century, developed nations' CO_2 emissions were unsustainable. Rich countries could get away with increasing carbon emissions only because poor countries had lower emissions. Of course, now the atmosphere is already degraded, and everyone is paying a penalty, though in unequal ways. It seems reasonable to say that if a community cannot universalize its practices, then it is acting unsustainably.

The picture that emerges from using this model is a nested set of communities, from a few people in a place to bioregions, national communities, and finally to the global community. At each scale we can look at how the dynamics of the community affect its own and others' capital stocks and also how practices at larger and smaller scales are affecting the community we are interested in (our "focal" community). If the stocks of capital are high enough to permit continuing regeneration, and they are not decreasing, then the community is relatively sustainable in the absence of negative impacts from the larger-scale systems in which it is embedded. To put this account formally:

Community C is sustainable during a period t if:

1. the stocks of natural, social, human, and financial capital within C's control are stable or increasing over t;
2. continuing the activities of members of C for multiple generations beyond t will likely further maintain or increase each of these stocks; and
3. during t, C achieves a level of impact on each of these stocks such that if everyone had that impact during t, human society would have the capabilities to thrive for multiple human generations within the carrying capacity of the planet (all else being equal).

So far, this broad account of sustainability leaves open some important questions. What is the relation between sustainability and other important social goals like justice? Does sustainability require transformation of central aspects of our current system, such as global capitalism, or does it preserve most of our current system while avoiding crossing key thresholds? A virtue of the model I have sketched is that it allows for different answers to such questions while engaging people in a continuing search for more refined visions of sustainability. Still, I do need to provide my own tentative answers to such questions so the picture of sustainability has more detail.

It is tempting to see sustainability as an overarching goal, which includes addressing poverty, racial injustice, and income inequality, as well as avoiding planetary boundaries. The United Nations Sustainable Development Goals are broad enough to suggest this view. The argument that most negative features of our society degrade the stocks of social, economic, or ecological capital seems to justify this position. Yet there are some strong reasons for highlighting the tensions between goals like sustainability and social justice by identifying them separately. The Brundtland definition of sustainable development with which we started seems to emphasize justice for future generations, but this can be in tension with advancing justice now for current peoples. For example, communities in the Global South may need to burn their easily accessible fossil fuels in order to develop their economies in the absence of a great deal more aid from the Global North. It is not enough to say that all present people can meet their needs, when there are gross inequities in standards of living.

Julian Agyeman has been a powerful advocate for "just sustainabilities," a view that explicitly combines social justice and sustainability.[22] He argues that sustainability, as it is commonly interpreted, is an incomplete guide to community goal setting; we must highlight justice. He agrees that without achieving social justice goals, genuine sustainability cannot be reached; but unfortunately many in the sustainability community emphasize the environmental portions of sustainability and downplay justice. Agyeman's examples focus on eliminating food deserts, creating healthy human habitats in which vibrant urban neighborhoods support good jobs, and democratizing streetscapes so they serve everyone, not just those with access to cars. In just sustainabilities, such goals are as important as species preservation and carbon emissions reductions, and sometimes they need to take priority. Similarly, Scott Campbell argues it is important to highlight the tensions that urban planners encounter between stakeholders focused on social justice and on sustainability, rather than to pursue a premature synthesis of these goals.[23] I concur that it is best now for pragmatic reasons to see such goals as distinct though intertwined.

The second question—how much transformation, if any, sustainability requires—involves speculation that often reflects ideology more than empirical evidence. Many people pursue a vision of sustainability in which we appear to be trying to extend the conservation phase of modern postindustrial global capitalism and the lifestyles that it enables for the well-off while raising others toward such lifestyles. This vision is supported by the pursuit of technological fixes for planetary limits, such as renewable energy, carbon capture, genetic preservation of species, large-scale ecological restoration, and high-yield genetically engineered crops.

An alternative vision of sustainability sees our system and the lifestyles it enables as the causes of unsustainability and thus insists that we need to transform the identity of this system and reorganize in a more sustainable pattern.[24] The transition-town movement is a well-known example of the latter view, featuring major shifts in energy use and lifestyle leading to relocalization, re-skilling people, and more participatory governance.[25] In the debate about how much change is necessary for sustainability, I lean toward the transformational view. But of course, we cannot know in any detail which elements of the system can be preserved in the process and what should be lost amid reorganization. We can be pretty sure that sustainability will require fundamental changes in energy use, but we cannot be sure whether use of personal vehicles will decrease along with fossil fuel use or whether electric vehicles will enable a more sustainable world that retains the freedom of the road.

The either/or implication of the transformation-or-conservation question is misleading. Sustainability will inevitably require some adaptive cycles on many scales to enter release and reorganization, leading to changes that lose elements of our identity we would prefer to keep. But unless we enter major collapse socially, our progress toward sustainability will preserve many parts of our systems that are important to us. Thus, key elements of our identity will be conserved. Some sustainability advocates want the system to crash down and be replaced, and others want much of it to be preserved, but sustainability itself does not require either. Much depends on what behaviors we are willing to change on a large scale and what innovations—technological or political—we can develop fast enough. And there is much value in the opposing visions competing for our allegiance as the future unfolds.

The case for thinking that just sustainability spreads flourishing should be fairly evident by now. If on large scales we do approach net positive impact on social, environmental, and economic stocks and ensuring they are at a level sufficient for system functioning, then on these scales the following are likely to be true: We will have limited our risk of crossing major planetary boundaries and suffering the chaos that comes with the release that would have followed. We will have developed governance norms than enable us to manage much conflict and work toward solutions to our challenges on multiple scales. Trust in institutions and between people with different views will be on the rise. We will have preserved enough species and well-functioning ecosystems that we have the ecological services on which we depend. We will have alleviated much poverty and made systemic progress toward just treatment of present and future peoples. A critical mass of people will believe that they are prepared to effectively navigate the dynamics of their times and contribute to the progress communities seem to be making. This list could be much longer, but this is enough to show why under such conditions, a great deal more people would be able to flourish. Most systemic barriers to flourishing would be diminished.

The above list is also tailored to echo points in the last four chapters that link skillful habits to addressing our challenges and grasping the opportunities that lie therein. Widely embraced frugality skills would contribute significantly to reducing negative ecological impacts and reducing unjust distribution of resources. Collaborative and humility skills will be crucial for reducing conflict and crafting governance that can address our challenges. Empathy skills would sensitize us to the plights of less fortunate peoples and other organisms. Systems skills would help us productively address our challenges and navigate the turbulence that is an inevitable concomitant of where we are on large-scale adaptive cycles.

The sheer number, scale, and complexity of the challenges we need to surmount to approach sustainability are daunting. As noted earlier, if these challenges are addressed one at a time, we will almost certainly fail to meet many of them. Sustainability requires us to look for key leverage points that enable us to address many challenges at once. We must find an approach that enables actions to be unified, easy to communicate, practical for ordinary people, and focused on addressing root causes, rather than symptoms. In the next two chapters, I will explain why cultivating resilience skillful habits is a particularly promising leverage point and how we might accomplish that through education broadly understood. By ourselves, most of us can do little to change our culture's trajectory to create a smoother and more rapid transition toward sustainability. But we can participate in social movements that promote such goals.

BARRIERS AND STEPLADDERS

Scaling up resilience skillful habits is reasonable only if the barriers to that project can be surmounted. I have addressed objections to specific skillful habits in prior chapters, but I still owe the skeptical reader a charitable discussion of issues with my whole approach to flourishing in our times. In fulfilling this obligation here, I aim to exhibit the skillful habits I am recommending. This requires framing the issues as genuine barriers others may encounter in contemplating my arguments, not as misguided objections to be summarily dismissed. My task, then, will be to provide stepladders that will help people climb over the barriers. This kind of intellectual "trial by fire" is a powerful way to show how a view is supposed to work in practice and to illustrate how its parts are interconnected and mutually reinforcing. In the process of crafting stepladders, I draw out key further implications of my proposals. The result will be a synthetic overview of how our odds of flourishing might be enhanced despite the barriers.

I begin with the barriers to individuals pursuing flourishing by cultivating resilience skillful habits. Some are theoretical barriers—critiques of the ideas themselves or reasoning I offered. Others are purely practical—concerns that some people will face barriers preventing them from cultivating resilience traits to an extent that will contribute to their flourishing. Yet others identify risks likely to outweigh any benefits from cultivating the skills. I have ordered the construction of the stepladders over these barriers to create a path that illuminates some of the more obscure corners of the project. The chapter ends with the barriers to scaling up cultivation of the skills. Most of these I acknowledge to be serious issues; my responses will be sketches of ladders that have an aspirational element, which the following chapter will try to make inspirational.

Individual Flourishing Hurdles

The elitism barrier: Perhaps the most common concern I have heard is that cultivation of these underdeveloped skills is not realistic for most people, so even if these skills did increase the chances of flourishing, it will only be for the few. Thus, the result is elitist. Three reasons appear to provide support for the elitist objection. First, swimming against the cultural current is easier for people who already have a fair amount of cultural capital; they can afford to be rebels and often can get away with it. Marginalized people often cannot afford that risk. Indeed, they have to try harder to fit the current cultural norms in order to avoid running afoul of institutional expectations. Second, cultivating underdeveloped character traits takes time and resources, which many people do not have after working a full-time job (or two), taking care of an extended family, dealing with health issues, and so on. Third and more controversially, the detailed skills I have sketched seem to involve nuanced judgment that few can master, even if they are highly educated. It is not realistic to think that regular people will be able to develop these skills to a great enough extent to make a difference in flourishing. In short, this is a privileged approach to flourishing, which only works for those who are already well situated to avoid the worst consequences of the age of climate change.

The practical challenges outlined above are real and important. Working on skills does take time, and challenging cultural norms comes with risks. At the same time, it is doubtful that on average the barriers for non-elites are higher than for elites. Indeed, sometimes the barrier is likely to be greater the more privileged one is.

Our capacities to acquire resilience skills vary considerably, depending on our personalities, our situations, and our inclinations. People from many walks of life already have well-developed versions of the skills, but most of us need to strengthen them significantly. Our capacity to learn skills depends on motivation levels, life experiences, the subcultures we inhabit, as well as time and resources. Often people who are members of non-elite groups have motivations, experiences, and subcultures that enable them to deepen the resilience skillful habits more rapidly than elites. Privilege can diminish the motivation to acquire such skills. This is most evident in the case of frugality and humility skills.

People who live through economic hard times often acquire frugality skills out of necessity. They can find role models nearby. By contrast, those who are wealthy have a harder time motivating the acquisition of such skills. They have less need for frugality and less experience that cultivates the skills. Indeed, conspicuous consumption has been rewarded for this group. It is unlikely that marginalized groups are worse off with respect to developing frugality skills. If anything, it is likely that the balance tips in the other direction.

Similar arguments can be made about humility. Those who are well educated and accustomed to wielding power are often overconfident about their own views. They are more likely to be concerned about status and entitlement and less focused on the limitations of their perspectives. On average, those who have not been empowered often have an easier time listening openly to others and seeing their points of view. Few studies compare humility across populations, but in one suggestive study, African Americans and Arab Americans in the Detroit area scored higher on a humility scale than a comparable white population.[1] It is far from clear that the powerful have an advantage with respect to acquiring humility skills.

Some collaborative and systems-thinking skills do appear to be easier for elites to cultivate. For example, systems skills are typically taught, if at all, in college. Thus, one can argue that elites have an advantage there. Skills associated with trusting are likely to be harder to develop for those who have been repeatedly disadvantaged by people in power. Indeed, robust evidence shows that trust in others tends to be lower in groups with lower socioeconomic status.[2]

Empathy skills may not track trust skills, though. In one study, Michael Kraus and his colleagues tested the hypothesis that people in lower socioeconomic strata are more dependent on others in their group for support, and thus that they developed strong skills of cognitive empathy.[3] They found that indeed individuals from lower-class backgrounds made more accurate empathetic judgments than people from upper-class backgrounds. They found some evidence that sensitivity to social context was a likely cause of this variance.

We definitely need more research on variables affecting acquisition of skillful habits. But both experience and our very limited data set suggest that differentials between elite and non-elite groups in the resilience skills are likely to be mixed at best. In the absence of evidence to the contrary, it seems reasonable to conclude that the barriers to acquiring resilience skills are not markedly different for the two groups. This is not surprising, given that most elites have learned how to play the culturally rewarded games well; that is, they tend to excel at skills reinforced by dominant norms. They may also excel at contrasting skill sets, but they often lack the motivation to learn them.

This leaves the third reason one might think elites have an edge in acquiring sophisticated resilience skills, namely that acquiring these requires a degree of nuanced judgment that is more likely to be mastered by groups that are highly educated. Perhaps anyone can acquire these skills to a limited extent, but for the skills to increase the likelihood of flourishing significantly, they must be well developed and integrated into effective habits. I would agree that the stronger the skills and more habitually used, the more effective they would be in enhancing flourishing. But I am not aware of any evidence for thinking that education is well correlated with the nuanced judgments involved in such life skills.

The research on cultivation of wisdom addresses how we develop our capacities for good judgment. In this literature, wisdom is distinguished from abstract reasoning capacity.[4] Formal education does try to enhance reasoning and critical thinking, but it rarely focuses on developing context-sensitive judgments about how to balance conflicting goods that help refine skills like humility, empathy, frugality, and the like. Given the role of life experience in strengthening such judgment, it seems unlikely that elites have an edge here. Few studies have looked at differences in wise-reasoning capacity between demographic groups, but a study by Brienza and Grossmann found that on average people in lower socioeconomic status (SES) groups have stronger wise-reasoning capacities regarding interpersonal relations than people in higher SES groups.[5] They suggest that this could be explained by the importance of fruitfully navigating relationships when resources are limited and other threats abound.

Individuals with strong resilience traits can be found among any group. I encourage readers to think of their own examples. One salient example for me was Carl Stoddard, a beloved security guard at Green Mountain College for almost thirty years. When he died in 2004, students and alumni from around the country told their stories about Carl and funded the creation of Carl's Corner, a small monument where Carl had leaned on a wall as he watched over the college most nights. Carl had a high school education and grew up amid the rough-and-tumble slate miners around Granville, New York. Students loved him for his humor and compassion but also for his good judgment. They would seek him out to get advice on everything from relationships to growing vegetables. Security guards must find a balance between upholding campus rules and making judgments about what is in students' interests. Carl was not shy about making such judgments and defending them before administrators who might be more rule-focused. He was not lenient; he just knew how to get people to behave when it mattered most. He read people well, knowing when they needed a hug and when they needed professional help. Perhaps more than anything else, Carl was loved because he could help students frame a problem differently so it was more easily solved. He had the kind of nuanced judgment that is required for strong empathy, humility, frugality, and systems skills.

The barriers to developing resilience skills are unlikely to vary in a systematic way between elite and non-elite groups, but they do vary widely. Lack of motivation and suitable life experiences, lack of role models and supportive friends can be barriers. We have already encountered stepladders for surmounting these. Like Benjamin Franklin, we need to integrate practice and reflection on the skills into activities we do as a matter of routine. We need to start where we have some motivation and experience, refining some skills we already have to some extent, and then building outward. We can create groups that support skills development in the process of doing other things that are fun or practical. Such groups can provide low-risk settings to explore skills that contrast with dominant norms. And we can work on the binocular vision that enables us to shift skillful habits as the context demands, avoiding risks of swimming against too strong a social current.

The "it depends on circumstances" barrier: The elitism concern suggests a deeper question about whether local circumstances play a strong role in determining which skillful habits tend to promote flourishing. In our increasingly diverse and pluralistic culture, some may be skeptical that it is possible to defend substantive generalizations about the broad skill clusters that promote flourishing. Just as situationists doubt that there are stable traits at all, these critics think that even if there are some stable traits, we would need to vary our emphasis on them depending on the circumstances.

No doubt having some degree of flexibility in the skillful habits we express in a context is important. That is a key reason why developing binocular vision regarding the contrasting skillful habits is important. Still, we can generalize across contexts regarding which traits are underemphasized and important for the challenges that affect all of us regardless of circumstances. The planetary boundaries we are approaching affect everyone, as do polarization, gridlock, and inequality. We would all benefit from finding jointly acceptable ways to make progress on these difficult issues. The skills we need to emphasize to make such progress would also benefit us individually in myriad ways as we navigate more local issues.

The stepladder we need here is to detach from our local focus and look at the larger systemic structures that will affect our flourishing. Organizations must do this as a matter of routine when they plan strategy: they must analyze trends at larger scales to decide which opportunities to pursue and risks to avoid. We should do the same. If we expand the scale of our vision for the purpose of planning and remove any ideological blinders that narrow our range of vision, most of us will find the interconnected challenges and opportunities that constitute the age of climate change. We may differ regarding specific approaches to addressing these challenges, but it will be easy to see that some skills are largely missing from our toolbox. How we would develop and use these would naturally vary according to circumstance.

The question of how far resilience traits can be applicable across contexts cannot be settled here. A strong case can be made that some traits are widely applicable, such as being honest with ourselves and our close companions. Though even here, with a bit of effort we can imagine situations where some contrasting traits (e.g., misdirection and obfuscation) are equally conducive to flourishing. Imagine a world in which most human interactions have the structure of a poker game. What we are looking for is traits that are conducive to flourishing across a wide range of circumstances in which we are likely to find ourselves. My position is that the traits associated with flourishing shift depending on where a culture is on large-scale adaptive cycles. Thus, I share a concern about generalizing about the applicability of traits to all circumstances; yet I resist the claim that no broad generalizations are reasonable at all.

The "flourishing now is immoral" barrier: Even if flourishing is possible during our times, it may be that we *should* not try to flourish as we destroy huge numbers of ecosystems and cause untold damage to human communities. Perhaps it would be intensely self-centered to pursue flourishing while contributing to massive global disruptions. Indeed, it could be argued that a profound lack of empathy would be required to flourish while our lifestyles are causing so much harm. If we are sensitive to the amount of pain that will surround us as climate disruption, loss of biodiversity, and deepening climate injustice accelerates, we should regularly feel anguish, and this seems incompatible with flourishing. A good person would feel consistently very bad about what is happening in the world. Anyone who flourished in this context would be a deeply flawed person.[6]

The concern here is reasonable. People should feel grief and anger when innocent people and other creatures are being harmed. It is a sign that someone is empathetic and has a sense of justice. It may seem natural to extend this point to feeling constant anguish when the harm is prolonged and systematic. But this second move is neither necessary nor morally required. Indeed, even in the context of climate catastrophe, constant anguish would not be wise.

The stepladder needed to surmount this barrier is the skill of allocating attention and emotional response to objects of grief and anger to the degree that it helps us to learn viscerally about the consequences of our actions, to motivate shifts in behavior, and to feel deeply enough about a harm that we can then psychologically move on to the rest of our lives. We may have a hard time developing the skillful habits that manage attention in this way, but over time deep deficits in this area may become mental health issues, not rational responses to crisis.

Continuous anguish, anger, and despair are not compatible with flourishing, but powerful episodic grief and anger are compatible. Indeed, it is hard to imagine flourishing without having such emotions, because of the ways they inform us and shape our future behavior. Remember that flourishing is not just happiness.

Flourishing includes positive emotions, but it also involves other less subjective elements like close relationships and achievement. Someone may flourish without being happy much of the time. A person can flourish even if she experiences a great deal of pain—for example if her house burns down or she gets a terrible disease. But if this is so, then shouldn't someone else be able to flourish while being acutely aware of that pain? The key is to avoid so much focus on the harm that it prevents us from accomplishing other meaningful goals and appreciating other gifts.

Of course, if we are a cause of the harm, then we should feel guilt, and we should try to shift our behavior accordingly. For example, if we are not trying to significantly reduce the indirect harm we are causing to others through our carbon-intensive lifestyles, then we are arguably behaving immorally. But cultivating the resilience skillful habits does aim to reduce such harm. Frugality skills most directly enable us to significantly reduce our negative environmental impacts. Systems, humility, and collaborative skills enable us to work effectively in groups to reduce harms. If most people cultivated such skills and thereby shifted cultural norms, we could more rapidly advance the just sustainability transition. Thus, attempting to flourish through cultivating resilience skills should reduce guilt while increasing empathy for those most burdened by injustice and unsustainability.

The stepladders we need to surmount this barrier involve developing an acute awareness of ways we cause harm to humans and nonhumans and making a serious attempt to reduce our own contributions to such harm. Sometimes we must recognize that harm results from all of the options we face, and thus that our choices are tragic. Guilt is appropriate, where purity is impossible. Fortunately, humans can flourish in a deeply tragic world.[7]

The "progressives' values" barrier: Some will be concerned that I have misidentified the skillful habits that will help us to flourish in our times and that the skills currently dominant in US culture will actually be more useful than those I propose. This might be conjoined with the view that resilience skillful habits reflect progressive values, not a clear-eyed analysis of what flourishing requires. One version of this concern is that our dominant skillful habits are associated with "traditional masculinity"—habits like being able to compete aggressively, to act independently, and to command with conviction. Josh Hawley, in a keynote speech at the National Conservatism Conference in 2021, argued that the Left was trying to redefine traditional masculinity as toxic masculinity and to erode the virtues of *"courage, and independence, and assertiveness."*[8] The concern that the traits I recommend are anti-masculine or anti-conservative would be a barrier to a significant portion of the US population pursuing the culture change project. This would also suggest that polarization would make scaling up resilience skills into social norms unlikely.

The ideologies associated with individualism, competition, and conviction have been criticized by feminists and others on the political Left, which may be the primary source of this concern. I doubt, however, that resilience skills are consistently aligned with traditional gender roles, despite the stereotypes that are sometimes advanced. Indeed, many of these skills are neutral with respect to gender roles and political orientations, and others are quite mixed in their associations.

The systems-thinking skills seem gender and politically neutral. Although persistent myths attribute different cognitive aptitudes to men and women, extensive research shows this to be mistaken. For example, the view that men are better at quantitative thinking has been debunked by several large meta-analyses.[9] Frugality skills seem similarly neutral. For example, the increased self-sufficiency that comes from reuse, repair, and repurposing skills seems neither masculine or feminine. Among collaborative skills, hope and trust seem neutral, though the hopeful ability to supercharge motivation under high-risk conditions is sometimes associated with traditional masculine roles. Affective empathy has been associated with traditional feminine roles, but cognitive empathy skills (the ability to read people's emotions from behavior) are less clearly gendered, though there is some evidence that women have a small advantage in this area.[10] Humility skills also seem mixed with respect to gender association. Decentering self is often attributed to feminine roles, though reflective skills are rarely seen to be gendered. Interestingly, unreflective conviction based on gut reactions seems associated with both masculine and feminine roles.

No doubt attempts to systematically cultivate some resilience skills would be politically polarizing. We have learned recently that almost anything can be politicized. The best response is to note that the whole package of skillful habits is largely neutral; any political associations are quite mixed. Nevertheless, with some audiences it would be unwise to begin a defense of resilience skills with critiques of contrasting skills that are quite politically charged (e.g., competitive skills). Concern about polarization is not by itself a reason for limiting our skillful habit toolbox, especially when some of those skills could address polarization.

A different rationale for doubling down on our dominant skills is a picture of the world that emphasizes danger and intractable conflict. If we think the future will be totally conflict-ridden, then my emphasis on collaborative and humility traits becomes a more doubtful approach to flourishing. If we descend into regular violence as a means of addressing issues, we may need more highly developed competitive skills. Trust will become a luxury confined to close friendships, and empathy broadly expressed will be a dangerous distraction from the need for self-defense. For some today, this picture characterizes the only world they know;

that situation could become more common. If it does, then I have emphasized the wrong skills.

I have already argued against apocalyptic environmentalism and the level of conflict it suggests. If we believe the worst is on the way, it is more likely to occur, because we will not be able to act effectively given that pessimism. We cannot afford to be Pollyannas about adversaries we may face or future challenges, nor can we afford to be pessimistic Eeyores. We must develop the skills that enable us to hold firm between these two extremes. Some forms of release are probable, but the buffers we are strengthening will soften the impact of release, and our current pursuit of transformation is likely to accelerate the process of reorganization. No doubt some people will prefer to escalate violence rather than pursue transformations toward just sustainability; but for most that will be a last resort. Virulent competition for vanishing resources is a worst-case scenario, not a probable one.

The height of the above barriers depends on how caught up we are in the shrill fringes of contemporary political debates where crisis looms eternally and worst-case scenarios dominate our attention. For those in the large moderate middle, the stepladders useful here involve assembling a set of reminders about the structure of the flourishing project. First, we are not eliminating any skills from the toolbox. By further developing skills poorly represented in the toolbox, we would be adding capacity and flexibility. Furthermore, we can adjust our use of the skills in light of how well they are working in our contexts. Skillful habits are not mechanically activated and followed in a rote fashion. We must hone and adapt skillful habits in our local contexts while being mindful of the larger systems that influence these contexts. Second, we should focus on what the skillful habits achieve, not on their putative ideological associations. Start with skills that seem more inviting and useful and progress toward others that seem more foreign or ideologically vexed.

The "more important skills" barrier: Some critics may think that I have misidentified the most important skills for flourishing. They may agree that change of skills is crucial, but focus on other skills. Daniel Lerch in "The Six Foundations for Building Community Resilience" argues that courage is a crucial virtue we need to strengthen now.[11] We need courage to face our problems directly and to take responsibility for them. We need courage to collaborate with those with whom we deeply disagree. We need courage to try novel solutions amid great uncertainty. His explanation of courage includes the skills associated with persisting when challenges seem overwhelming. Angela Duckworth's "grit"—passion and perseverance—is such a skillful habit.[12] Grit is often celebrated as a key to success in daunting enterprises. An increasing number of schools now try to cultivate grit. Hope overlaps considerably with grit and courage.

Adaptability is another skillful habit that might well deserve to be added to our list.[13] Consider the following reasoning for its inclusion in resilience skills. It is naturally contrasted with a cluster of habits that are expressed in our attempts to control our environments. When our desires are not satisfied by the way the world is, we can either try to change the world or change our desires. The control orientation that dominates our culture fosters skillful habits that enable us to exert enough control over our environments that our desires can be satisfied. This orientation reinforces our emphasis on technological development skills that enable us to effectively manipulate the environment rather than on psychological skills that enable us to modify what we want. It has also encouraged prioritizing knowledge gained from controlled experiments, rather than knowledge associated with uncontrolled observation.

Despite the benefits of control skills, prioritizing them has costs that are not broadly appreciated. Many important kinds of knowledge, like the wisdom that comes with life experience, are not susceptible to controlled experiment, yet we cannot avoid making judgments that require such wisdom. We must sharpen our abilities to evaluate better and worse judgments without the benefit of control. Though we can modify much of the world, control is often an illusion, especially on larger scales. Human behavior is particularly difficult to control. In the late conservation phase of the adaptive cycle, the brittleness of control regimes reduces resilience and eventually leads to release. We need more adaptive management to prolong the conservation phase, but if release comes, we especially need the flexibility to survive the increased chaos and begin the process of reorganization. Rapid experimentation is crucial in this phase. We must become much better at adapting to what we cannot control. As Montaigne said, "Not being able to govern events I govern myself, and if they will not adapt to me then I adapt to them."[14] We need both adaptability and control skills, along with the kind of binocular vision that enables us to use them in concert.

Adaptability, courage, and grit *will* be important during this age; they are involved in resilience, and they are currently underdeveloped. I make no claim that the resilience skillful habits constitute the only underdeveloped traits necessary for flourishing in our times. I have chosen the four key resilience skillful habits for two related reasons. For each of them there is a large literature that justifies their importance for addressing our major challenges, and each skillful habit includes numerous important subskills that will assist us in flourishing. The resilience skills clusters include under their umbrellas skills that might have been placed within a different cluster or made into a major category. Alternate ways of clustering skillful habits may be more intuitive or useful for different people. I have described hope skills as a subset of collaborative skills, but others might well elevate hope to a major resilience trait and include grit, courage, and creative reframing of goals among the clusters of skills that contribute to hoping well.

Given that the core idea of this book is that we are unlikely to flourish unless we change ourselves, it might seem that adaptability is the core resilience trait. Indeed, I could have packaged many resilience skills under the general heading of adaptability, but I doubt a primary focus on adaptability skills would motivate development of the necessary range of specific skillful habits. A separate chapter on adaptability would overlap too much with the other chapters. We should not seek a unique best way of describing the skillful habits that will enhance our flourishing in these times. A degree of open-mindedness, pluralism, and flexibility about the framework for categorizing resilience skills provides the most promising stepladder for surmounting the other skills barrier.

The "no silver bullet" barrier: We may be disappointed if we try cultivating resilience skills and flourishing does not seem to follow quickly. The stepladder here is increasing clarity about the connection between character and flourishing. Strong resilience skillful habits *improve our capacity* to flourish in our times. They are one important contributor to flourishing, but there are others, some of which are beyond our control.

We need to keep in mind the objective list account of flourishing I sketched in the introduction, which includes the following: positive emotions, engagement, positive relationships, meaning, accomplishments, having basic needs met, having an understanding of the forces shaping one's life, and reasonable confidence in one's capacity to navigate those forces. Resilience traits directly influence some of these dimensions of flourishing, and act indirectly on others. Here are some of the connections we have explored: Systems-thinking skills increase our understanding of what is happening on multiple scales and provide insight into how we can effectively navigate the personal challenges that result from impacts at larger scales. They build justified confidence in our ability to adapt in a turbulent world. Collaborative skills help us to build stronger interpersonal relationships and to work effectively with different groups to achieve larger goals. Humility skills reinforce the positive impacts of systems thinking and collaborative skills and in particular help us to avoid myopic approaches to problem-solving resulting from hard convictions ill-suited to navigating a rapidly shifting world. Humility skills also help strengthen relationships by decentering the self and enhancing curiosity. Frugality skills increase our chances of having basic needs met and accomplishing much within the context of constrained resources. Together, the above impacts of resilience skills are also likely to enhance positive emotions and increase our sense that our lives have meaning in a world where much is going badly wrong.

Our situations, our resources, our past experiences, our health, our other skills, and our social networks, among many other things, affect our capacity to flourish. Trauma of many sorts makes it harder to flourish, but those who overcome trauma are often better situated to flourish in these times than those who have been too sheltered. The Stoics argued that if you develop the right character, you can flourish in any setting; I make no such claim. But I have argued that skillful habits play a crucial role in enabling us to make the best of the forces that are shaping our lives, to grasp the opportunities that lie within the challenges, and to acquire what we need to flourish now.

The loneliness barrier: For almost half of my graduate students, the sense of isolation in doing the work of moving toward sustainability is a significant barrier both to progress and to their flourishing. Many live in places where they find little acknowledgment of the challenges we face and little support for the changes required to meet them. The loneliness of their pursuit often makes their work less productive and deprives them of the joy of joining with others to make change. They have to create their own positive reinforcement strategies in order to fuel motivation, and often they find it easier to focus on other issues. Fortunately, other students find themselves riding a wave of social engagement in addressing injustices, climate change, waste reduction, and so on. They also find positive reinforcement and role models for the development of many resilience traits. There is no easy stepladder for the barrier of loneliness. We either find communities of support or we have to create them. This project is the first step in scaling up cultivation of the resilience traits.

Scaling up is not just an answer to loneliness—it is a requirement for promoting flourishing on a wide scale and beginning to approach sustainability. Scaling up can occur in many ways, but it typically begins in small groups that have come together in order to promote some goals that individuals alone cannot achieve as effectively. If successful, a group may have enough influence on a small place-based community or organization that it develops social norms that reinforce the ends. Sometimes leaders facilitate this, but often it occurs in a bottom-up fashion where an increasing number of people are attracted to the group without a clear leadership structure. When this occurs in geographically dispersed communities, we begin to talk in terms of a social movement that aims to change the culture at large. If conditions are ripe, key organizations push the goals at the cultural scale, and major institutions may provide support for spreading the norms. If a social movement is successful, the culture itself changes, leaving inevitable pockets of resistance fighting uphill battles to retain the hegemony of the earlier culture.

Barriers to Scaling Up toward Sustainability

The scaling-up process is happening, though not fast enough, to be sure. Despite the mainstream media's gloomy tilt, we are seeing a seismic shift in the prevalence of practices associated with social justice and sustainability. Education and business increasingly play a leading role, but much of the impetus started with ordinary people trying to do something differently on a small scale. The nonprofit sector has been crucial in aggregating and accelerating solutions to issues and creating a larger audience inclined to action. This movement focuses primarily on overt behavioral practices, but it also involves the cultivation of resilience skillful habits. A fierce politically motivated resistance has expanded the culture wars to include this movement and succeeded in polarizing some proposed solutions. Though a just sustainability still seems a distant dream, it is becoming easier to find experiments in social norm shifting that are succeeding on smaller scales.

Paul Hawken's *Blessed Unrest* provides a powerful overview of "the social movement that has no name" that has been galvanizing people around environmental and justice issues.[15] More than sixty years of activism have illuminated the linkages between these issues. We have learned how empathy for marginalized others, human and nonhuman, can propel change, and how communities of hope can reinforce our motivation and skills for persisting amid determined opposition. To document this movement, Hawken and others at the Natural Capital Institute created a huge database of organizations in 243 countries and territories that are engaging communities in addressing humanitarian issues. Now dated, this time slice of the movement helps us to see vividly that we are scaling up culture change and in the process cultivating many of the skillful habits needed to approach just sustainability. Andreas Karelas provides a much more recent, though narrower, account of the advances in the climate change movement in *Climate Courage*.[16] Both books powerfully illustrate how movements can scale up culture change. But significant barriers impede scaling up the cultivation of resilience traits.

The "culture change takes too long" barrier: After decades of inadequate action, we must now move very quickly to address climate change if we are to limit harm to manageable levels. Biodiversity loss, income inequality, racial injustice, and polarization have similar urgency. Some will be concerned that the process of scaling up resilience skillful habits to initiate a shift in cultural norms would probably take multiple generations. On this view, cultural change may piggyback on other sustainability initiatives, but it cannot be a primary driver of the sustainability transformation because of the need for speed.

We do need to act as quickly as possible, but we should not adopt quick fixes that fail to address underlying root causes. In the long term that will be less effective. Unfortunately, our cultural norms often do evolve slowly. This barrier is considerable, but we must find ways to climb over it, because we are very unlikely to be able to go around it by using other levers of change. We must find ways to accelerate culture change, and we must integrate such change into other actions that will have more immediate positive impact.

Just encouraging individuals to strengthen the resilience skillful habits is inadequate for accelerating cultural change; it depends on the action of too many diverse individuals. That would be a bit like trying to solve the litter problem by encouraging more people to pick up litter. It is very hard to leverage system change simply by aggregating individual actions. We need a social institution that can leverage change in resilience skillful habits on a massive scale. Education could do that.

Formal K–12 and secondary education have historically focused on cultivating skillful habits, communicating knowledge, and transmitting cultural norms to new generations. They have often played a key role in accelerating the spread of norm changes. Although much education has reflected the individualist and competitive norms of our larger culture, we have seen a groundswell of educational innovation over the last twenty years that is moving in a different direction. In the next chapter, I will provide an overview of such innovations and describe how education can be transformed to accelerate culture change toward an emphasis on resilience traits. Formal education is unlikely to play this role by itself; business, religion, the family, and the state must play supportive roles as vehicles for education broadly understood. Admittedly, achieving the speed of change necessary is a long shot. The barriers to leveraging change through education, which I sketch below, are considerable. But trying to surmount them is our best shot at addressing a root cause of our current challenges.

We have good evidence that shifts in cultural norms sometimes happen fairly rapidly. Often this occurs when some crisis forces the change. We saw that the Great Depression and the austerity that accompanied two world wars played a key role in motivating the return of many frugality norms in the early twentieth century. Arguably the COVID-19 pandemic has shifted norms regarding telecommuting for office workers and tele-health in the medical community. Education has played a key role in accelerating change in norms around use of technology and acceptance of diversity in the last thirty years. Even a deep cultural norm change regarding individualism has occurred fairly rapidly if Putnam and Garrett are right.[17] Given the fractured nature of culture in the United States, we will need a focused effort and an effective lever to shift the emphasis from currently dominant norms to norms that reinforce resilience traits.

One stepladder for surmounting this "urgency" barrier is integrating use of resilience skills into short-term initiatives that make progress on our challenges. We should pursue policy and technological solutions in ways that complement culture change. We need rapid and sustained investment, incentives, and policy changes to accelerate the shift to renewable energy. We need to shift tax structures to address economic inequality. We need changes in policy and enforcement to reduce racial injustice. These along with many other immediate initiatives are necessary but far from sufficient to move us toward authentic sustainability.

Sometimes the tactics used by Hawken's "movement that has no name" do not reinforce the resilience traits we need to cultivate. It is tempting to use the dominant cultural norms to achieve the above initiatives even if one supports alternative norms; this often appears to be the most rapid path to success. For example, the conviction conveyor belt is an attractive tool to galvanize supporters, as is formulating options in terms of binary choices that force people into opposing camps with competitive orientations. Sometimes complex system analysis gives way to solution slogans that obscure the deeper leverage points for change. This tactic is understandable because of the need to mobilize lots of people who do not understand the system dynamics, but it often makes forward progress harder.

As social movements intersect with politics, they become associated with the use of hyperbole and the tactic of demonizing opponents. This reinforces competitive skills and makes collaboration unattractive. The end result tends to be an appeal to force rather than the search for reasonable compromise, not just among politicians but also among many in the movements. Collaborative and humility skillful habits seem to signal weakness rather than wisdom in such contexts. Our mainstream media tend to amplify the appearance that movements aligned with both political parties are engaged in blood sport, which obscures many local collaborative achievements and pushes competition down into local arenas. If we pursue policy and technology changes by using the dominant norms that reflect one of the root causes of our unsustainability, we will impede the shift in cultural norms and may produce more short-term fixes that backfire.

Then how can we avoid competitive approaches if a determined opposition seeks to derail all short-term progress? It is very tempting to try to beat the opposition at its own game—countering mistaken hard convictions with our own hardened convictions, matching unfair competitive tactics with similar responses, and using force to avoid being forced in the wrong direction. Some of this is inevitable, but if it becomes the dominant response to opposition, then it reinforces skillful habits that make collaborative solutions impossible. It is foolish to try to collaborate with those who refuse to engage, but it is wise to use a two-pronged strategy, always seeking to create common ground, while steadfastly deterring aggressive responses. This is a balancing act that often makes one unpopular with both sides, who often think they have the strength to win a fight.

The polarization barrier: Many will worry that political polarization makes widespread norm change seem like a naive fantasy. Individual freedom and competitive politics are core elements around which polarization turns. We cannot expect the set of resilience skillful habits to avoid politicization. Even within this context, we can still work on strengthening resilience skills among portions of the population open to the enterprise. This positions us for norm shifts as conditions change. But the use of formal education as a lever for change becomes more fraught if it is seen as being used to inculcate the habits associated with one side of a politicized divide, as we have seen with concerns about teaching "critical race theory" in the schools.

This barrier to the use of education to accelerate culture change is potentially very high. Yet some stepladders can help us surmount it. First, focusing on skills development in education rather than on dispositions or beliefs that have ideological associations is much less likely to polarize. Teaching students how to collaborate effectively is not as easy to stigmatize as teaching them that competitive approaches to social problems have important limitations during the age of climate change. Collaborative and systems skills are demonstrably important for a wide range of careers, and they can be taught without reference to social critique. Humility and frugality skills may invite more critique, but again, these can be detached from perceived ideological viewpoints. Second, it is important to keep salient that the aim is greater binocular vision regarding contrasting sets of skills. We are not aiming to eliminate individualistic, competitive, or conviction skills. We are broadening the toolboxes that students can use in their own pursuit of flourishing.

Character education has not yet been highly polarized, even though it is making a comeback after almost a century of neglect. People on the left and the right have strongly supported an increase in character education, and the results have included a range of character traits that dovetail with our resilience traits. For example, Texas is implementing new statewide standards for character education, which include empathy, compassion, concern for the common good and community, fairness, freedom from prejudice, perseverance, self-control, and gratitude, many of which are elements of resilience traits.[18] Texas includes a range of other traits in its list. It is too early to tell whether these standards will be implemented in ways that avoid polarization, but it is promising that the character education movement is both growing and avoiding polarization.

Inevitably, we will have debates about which skillful habits are most important to cultivate and which methods of cultivation should be used. These can be healthy and fruitful for a community. We should not expect agreement about answers, and we should be ready to learn from the debates. We will need an upswell of support for educational changes that focus on broad skills development, not just reading, writing, and mathematics.

The "risks to good education" barrier:[19] Even if we can avoid polarization when using education to scale up cultivation of resilience traits, other barriers result from risks associated with problematic implementation of character education. Three risks are particularly troublesome. First, character education could violate the autonomy of students and their families by imposing acquisition of skillful habits against their wishes. Second, it could also result in the homogenization of a school's student body and thereby reduce kinds of diversity that contribute to quality learning. Lastly, it could lead to undue moralization of common skills and a culture of shaming those who manifest contrasting skills. It is worth expanding on these potential issues here and indicating in general how these risks can be minimized. The real test of the requisite stepladders will be in the next chapter, where we will look at how education should change in order to cultivate resilience traits successfully. We will see that if curricula and pedagogy reflect the resilience traits they aim to cultivate, the risks will be manageable.

The first concern is that public schools will impose controversial skillful habits on students. The curriculum might indoctrinate students with views that support the skills, and students could be required to manifest such skills in order to progress. Many families have no choice other than the public schools; they would have a reasonable concern if nonstandard skillful habits are being required. Few would question imposing academic honesty or respect for community members in this way. In a Catholic school, one might expect some Catholic virtues to be cultivated in this way. But a school courts the charge of violating autonomy if it requires the cultivation of a skillful habit that is not widely accepted as being important for being morally decent, or being educated, or being a good citizen. Arguably most of the traits on the Texas list fit within these categories, and so do many resilience traits. But what about gratitude? Should a public school require that someone demonstrate gratitude sufficiently in order to graduate from high school? Alas, many teenagers would flunk out, and I would have flunked too. Requiring that graduates can write well seems much less invasive than requiring that they be appropriately grateful.

The autonomy barrier is constructed by raising concerns about imposing, indoctrinating, and requiring controversial skills. The problem here is the approach to teaching, not the content. The barrier shrinks considerably if a school seeks only to expose students to resilience skills, to highlight their uses, and to practice them enough for students to make an educated choice about whether to internalize them. The students can then choose whether to ignore them or make them into lasting habits that are self-defining (or something in between). In this approach, the school does not require such a choice for progress. Now the risks are minimal. Even if resilience skills were explicitly endorsed as norms at a school, this approach to cultivating them still avoids indoctrination. In most schools, contrasting skills would also be cultivated and represented in a diverse population of faculty and students, since they reflect dominant norms in the larger culture. In such contexts, the charge of violating autonomy seems specious.

The threat of homogenization emerges anytime a school builds a strong culture around its values. Suppose a school places a high premium on collaborative approaches to problem-solving, recruits students who are interested in learning the requisite skills, and downplays competitive skills. Imagine that it employs pedagogies that reinforce collaborative approaches to learning, and it celebrates its collaborative cultural norms. Over time, the school may come to lack diversity in approaches to social problem-solving within its community, which could decrease the scope of students' skills toolboxes to their detriment. Strong school cultures tend to increase student learning, but schools should avoid the loss of diversity that can accompany such strong cultures. Homogenization verged on being a problem at Green Mountain College; in the next chapter, I will suggest how that problem could have been better addressed.

Fortunately, the content of the resilience traits supports the kind of balanced curriculum that promotes productive diversity. Humility skills require us to recognize the limitations of any single approach to social problem-solving, including collaborative approaches. They encourage familiarity with a diversity of skillful habits. Collaborative skills are best learned in a context where there is enough diversity to make collaborative processes difficult. Someone who is good at collaborating must know how to work with others who lead with competitive skills. They must be able to respect such people in order to effectively build collaborative relationships. Furthermore, to deeply understand the dynamics of the systems we inhabit, we must understand how different skillful habits shape people's interactions, which is best done through experiencing such systems on smaller scales within a school's learning community. The resilience skills serve as guardrails that prevent the homogeneity that critics fear and require diversity for their own cultivation.

The third risk applies to all educational programs that have a strong normative core. They can foster a sense of moral superiority in those who share the norms taught in the program, which can lead to shaming others who violate those norms. A school with a strong sustainability culture risks graduating students inclined to shame those who drive "gas-guzzling" pickup trucks, wear sneakers made in sweatshops, or enjoy fast-food burgers. We understandably resent the intrusion of judgments that imply the superiority of other lifestyles. This kind of moralization divides people and renders collaboration across groups more difficult.

Such feelings of moral superiority are incompatible with the full set of resilience traits, as is the practice of shaming. Humility requires one to acknowledge that we all fall short on some important skillful habits. Thus, recognizing a flaw in someone else does not warrant a sense of superiority. Also, since people may interpret or prioritize traits differently because of their own backgrounds or contexts, we must temper moral judgments about others with the recognition of all that we do not know about their situation. A humble person ventures moral evaluation with the presumption of mutual respect and care. The collaborative skills reinforce this approach. Basic respect is crucial for collaboration, and the practice of shaming tends to diminish respect. Insofar as collaborative people aim to build bridges between diverse peoples, they should avoid feeling morally superior to others. This does not mean collaborative people will not judge others, but judgments will reflect genuine understanding and charitable interpretation. Where resilience traits are reflected in a school's culture, the risk of a censorious political correctness will be much diminished.

Each of the above risks would become a significant barrier to successfully scaling up resilience skills if the dangers are widely realized. Fortunately, the resilience traits themselves should minimize the likelihood of such dangers, as long as the methods of implementing resilience character education reflect the traits being cultivated. This approach to managing the risks requires us to keep in mind the dangers and to model the traits being cultivated. Good character education requires a series of balancing acts between protecting diversity and embodying strong norms, between emphasizing a set of skillful habits and requiring them, and between encouraging appropriate pride in students for their achievements in developing skillful habits and avoiding the moral superiority that leads to shaming.

Indeed, balance is a theme connecting the stepladders that help us to surmount a wide range of barriers to cultivating resilience traits. Balance has been a central element of virtue theory since Aristotle. It is embedded in his view that virtues are means between two extremes; for example, courage is a mean between rashness and timidity. Here a sense of balance is necessary to make good judgments about when we have achieved a mean between extremes. Valerie Tiberius argues that balance is itself a character trait that we have good reason to cultivate. She explains balance as "the capacity to appreciate the value of a variety of different projects or goals and the ability to focus on a project long enough to achieve the value that it has to offer."[20] Her focus is on balancing different projects in a good life, but she might also have talked about the balancing of skillful habits necessary to successfully engage in our projects. I agree balance does significantly enhance our flourishing. It is a crucial element in the resilience traits. It must guide their development and the process of scaling them up into social norms.

EDUCATION AND CULTURE CHANGE

Education is our most promising approach to cultivating underdeveloped resilience skillful habits on a massive scale.[1] To effectively leverage such change, education would need to shift both the skills it emphasizes and its own culture. Fortunately, such shifts are already occurring. Today we are surprisingly well situated to grasp the opportunity that education presents for scaling up resilience skills. First, several recent educational movements are providing strong support for moving along this trajectory. Second, leadership at multiple levels supports their development. For example, the United Nations, scientific advisory boards, national educational groups, and business leaders have urged development of strong collaborative and systems-thinking skills. Third, the pedagogies associated with developing resilience skillful habits have produced high overall academic performance. Thus, we have purely academic reasons for supporting education that cultivates these skills. Fourth, the motivation for learning resilience skills flows naturally from learning about how the systems on which we depend function, which is important for multiple education goals. Chapter 7 shows how we could build on current trends and overcome considerable challenges to develop a critical mass of people with resilience skillful habits. This would create the feedback loops necessary to shift cultural norms, which would in turn accelerate further development of the skills.

To be clear, I am not arguing that this result is likely or that there is an easy path to accomplish the goal. The inertia we would need to overcome is considerable, and resistance will continue to be fierce in some quarters. But with hope and persistence, we have a fair chance of making enough progress to begin the positive feedback cycle. Recall, we are not trying to remake the culture in a fundamentally different image. We are trying to shift the emphasis away from skillful habits that are ill-serving us toward contrasting skills that we need now, and to develop the kind of binocular vision that will enable us to use a mix of skillful habits when they best serve our needs. As we saw in the last chapter, education has recently accelerated shifts in cultural norms and associated skills.

We cannot avoid teaching cultural norms; we can only decide whether to do it deliberately or haphazardly. Recently we have seen a resurgence of interest among scientists in large-scale culture change motivated by social justice and sustainability issues. Major articles with titles like "A Framework for Intentional Cultural Change" and "Evolving the Future: Toward a Science of Intentional Change" have generated widespread discussion.[2] The maturation of some social sciences, as well as our increasing emphasis on interdisciplinary systems studies, has contributed to the view that intentional large-scale change is less risky than it once appeared. The mechanisms used to promote cultural change depend on the kind of change. Often social marketing is used to shift specific attitudes and behaviors, like recycling, tobacco use, or alcohol abuse. In shifting skills, however, we need guidance, practice, and evaluation of performance, which makes education, broadly understood, our best method for leveraging change.

When we think of education, we typically focus on formal, institution-based schooling—our pre-kindergarten through twelfth grade (PreK–12) and post-secondary system. Here an institution sets both the goals of learning and the means by which learning is conveyed. Because the early stages of this system affect almost everyone in the United States, it has high potential to influence the habits of the populace over time—but this means its impact has a time lag when we are trying to reach the whole population, which we can ill afford. Postsecondary formal education has less of a time lag, but it reaches a smaller portion of the population. When we move beyond formal education, we find a dizzying array of kinds of nonformal and informal education.[3]

In nonformal education, the learner selects the educational goals, but the institution determines the means by which these goals are met. These programs include continuing education courses, skills trainings, and after-school programs. Adult education avoids the time lag problem mentioned above, but since its programs cater primarily to those who already have an interest in a topic, they often do not reach a highly diverse audience. In informal education, there is no structured curriculum at all. The learner chooses the means for acquiring knowledge or skills, but an organization may produce the materials. Such materials may be sought out by a potential learner, or they may simply be stumbled upon. Because the information is often accessed by chance, informal education has a somewhat broader reach than nonformal education. Ultimately the distinctions between these different forms of education blur, but their potential for reaching different audiences makes all of them important for the widespread cultivation of resilience skills.

We know that family and early education play key roles in developing the cognitive and noncognitive skillful habits that impact life success.[4] Heckman and colleagues have shown that noncognitive skills like persistence, self-control, sociability, and curiosity play a key role in life outcomes, and that foundational skills in these areas tend to be learned very early, and more sophisticated versions of the skills are built on these foundations.[5] A number of the resilience skills we have explored involve these noncognitive skills, though systems thinking and some aspects of collaborative and humility skills are better thought of as cognitive skills. Schooling can remediate deficits in foundational noncognitive skills, but often not as efficiently as ensuring relevant early childhood experiences. One of the main arguments for universal preschool is to enhance education in such skills at an earlier phase in children's development.

Preschool and K–12 educational initiatives to strengthen social and emotional learning and reinforce character development have added to the suite of practices that could scale up resilience skills, especially collaboration skills. The sustainability and social justice movements have significantly broadened the use of some of these practices and added elements that increase the development of systems, humility, and frugality skills. These trends form a solid foundation for the deeper educational changes necessary to scale up resilience skills. We begin this chapter looking at formal education resilience practices that could be strengthened and made more broadly accessible and then turn to the question of how non-school-age audiences can be reached through nonformal and informal education.

Laying the Foundations

In Allyson Guida's fifth-grade classroom at Lakewood Elementary School in Sunnyvale, California, collaborative learning is a central part of the curriculum.[6] Students participate in team-building exercises at the beginning of the year and then do frequent group projects. In a project focused on learning about different global regions, groups of three students read an article and create a presentation on it for the rest of the class. The presentations are taped, and afterward the groups view and evaluate their own presentation using a rubric and then decide what they should work on to improve the next time. Such collaborative learning activities emphasize active listening, appreciating alternative perspectives, and conflict resolution, as well as the subject matter. Such practices are part of the social-emotional learning (SEL) curriculum and pedagogy that Lakewood and many other schools have adopted.

School systems that adopt an SEL framework systematically cultivate many of the collaborative skills I have described, especially the empathy skills. They also systematically cultivate some of the reflective and decentering-self skills I describe in the humility chapter, as well as the self-restraint skills in the frugality chapter. In earlier grades Lakewood students focus on developing a growth mindset and emotional regulation. In kindergarten, they practice going to the "Chillax Corner" when they are angry or upset, in order to calm down. The corner is a quiet place with stress balls and fidget spinners, where a student can choose to take a brief time-out to process some conflict before returning to classroom activities. The teacher positively reinforces learning from conflicts and mistakes. Students also practice taking "brain breaks" when they are having trouble concentrating, a brief time when everyone can get up, stretch, or jog in place, while answering a few questions about class content before settling back into a lesson.

CASEL, the Collaborative for Academic, Social, and Emotional Learning, identifies five broad areas of competence involved in SEL: self-awareness, self-management, social awareness, relationship skills, and responsible decision-making.[7] Each of these clusters of skills can be cultivated in age-appropriate ways across the PreK–12 curriculum. A 2011 meta-analysis of 213 studies on the implementation of SEL indicates that it significantly increases academic achievement and social skills over control groups, while reducing conduct problems and emotional distress.[8] Broad support is accumulating for wider adoption of SEL practices. In an interim report of the Aspen Institute's National Commission on Social, Emotional, and Academic Development, one of the commissioners, Tim Shriver, is quoted as saying, "We've arrived at a huge moment of leverage, and we need to seize the opportunity we have."[9] Nevertheless, despite strong bipartisan support for SEL, some people have tried to politicize its adoption and to reject texts that include SEL.[10] Their reasoning is strained, but such opposition may slow its spread.

Character education is another movement that is often linked to SEL, though it has a different orientation. Character education focuses more on core values and how these can become behavioral habits. SEL focuses more on skills. But these two approaches blend easily when applied in a school, and some organizations lump them together in the services they provide for schools. Both movements build on the tradition of teaching some social skills in elementary schools, but they are more systematic and research-based than prior curricula. Moreover, both advocate for sustaining their approaches as children advance through higher grades.

Character education has a long history in American schools, starting with the Bible-based education found in most American colonies. It has sometimes been understood as a form of moral education in which teachers tell students what they should and should not do, but today's versions are much less didactic and more focused on helping students think critically about who they might aspire to be. Many charter schools emphasize character development and link it explicitly with flourishing. For example, in the 2010s, the large network of KIPP charter schools made character education central to their mission and based portions of their curriculum on positive psychology's work on the relationship between character strengths and flourishing.[11] They cultivated traits like grit (passion and perseverance), self-control, social intelligence, optimism, and gratitude. Some state education standards are following suit, as we saw in the case of Texas.

The effectiveness of SEL and character education initiatives depends a great deal on how they are implemented.[12] Merely promulgating rules and indoctrinating about norms do not build skillful habits. For the autonomous development of any character traits, students need to have a sense of choice about who they want to be. They need to have the importance of character systematically reinforced, they need to have good role models, and they need lots of opportunities to practice skills, reflect on results, and receive feedback.[13] In addition, students need to see how the skillful habits they are learning are manifest throughout the school; these habits need to be reflected in the ethos of the place. Hypocrisy erodes even the most sensible lessons. Unfortunately, since most school personnel have excelled within institutions characterized by the dominant cultural norms of society, it is unusual for them to create an ethos that is in tension with those norms.

The social justice movement has had an increasingly large impact on schooling. For decades its emphasis on diversity and inclusion has influenced the collaborative skills taught in schools by emphasizing the skills necessary for genuine inclusion of those of different cultures, races, abilities, and sexual orientation. In the early grades, teachers focus on using culturally sensitive materials and examples, and they often raise issues of fairness. They also highlight the importance of empathizing with diverse peoples in order to collaborate effectively and demonstrate the value of diversity in shared deliberation. Insofar as social justice education helps us to be more aware of the ways in which our cultural biases influence our beliefs, it also enhances humility skills that involve understanding our limitations.

More recently, that movement has stimulated a greater emphasis on developing the systems skills associated with identifying and rectifying structural injustices; much of this work is done in the higher grades, where social studies and history classes lend themselves to discussions of the dynamics of power and the plight of marginalized groups. At this level, students can also hone their skills of assessing the different elements of fairness at play in social arrangements and learn how to use more nuance in negotiating conflicts around fairness. While many systems skills are taught only in college, high school students often learn the skills associated with developing individual agency to bring about social change and the importance of feeling empowered as a change agent. A wide range of leadership skills are tested and internalized in high schools if not before.

Most schools have explicit commitments to promoting diversity and inclusion. An increasing number also have programs that cultivate deeper understanding of the perspectives of marginalized peoples and the skills of assessing cultural bias. Relatively few schools make social justice a curricular theme across the school. One such school, the Capital Preparatory Harlem Charter School (grades six through twelve) uses social justice as a lens through which numerous subjects are taught. All graduates must complete a social justice project that identifies a problem, raises community awareness about it, analyzes data, and tries to address elements of the problem. Such project-based requirements are particularly important for skill building and empowerment.

The sustainability education movement augments the achievements of the above initiatives and has the greatest potential for scaling up cultivation of all four clusters of resilience skillful habits. Like social justice education, it has an emphasis on cultivating systems-thinking skills, and it increases the opportunities for building collaborative and humility skills. In addition, its practices include more emphasis on frugality skills. Sustainability education has been sweeping across schools in the United States, but it is best seen as a global movement. The United Nations recognized that education was fundamental to the goal of advancing sustainability and created the Decade for Education for Sustainable Development, 2005–2014, which had as its fundamental aim "reorientating education toward sustainability."[14] Global populations needed to learn how to think in new ways in order to develop economically while fostering more just social arrangements without exceeding our planetary boundaries. Various forms of environmental and sustainability education had been developing long before that decade, but the UN succeeded in moving these from the curricular fringe toward the center of education policy in many countries. In US primary and secondary schooling, this encouraged more states to revise educational standards across all grades and shift pedagogical approaches accordingly.

High-quality sustainability education often uses applied project-based bioregional curricula that complement SEL/character-education work on collaborative skills. Teachers often add place-based experiential components to a curriculum to build greater understanding of local system dynamics and to empower students to contribute to their communities. For example, according to the 2019 US

Department of Agriculture's farm-to-school census, more than twelve thousand schools in the US have school gardens that produce food and serve as learning laboratories. The number has grown significantly compared with prior census reports; in 2013 approximately seven thousand schools had edible gardens. Such gardens can be used to teach about the science of plant growth, the impact of climate cycles, soil science, and nutrition, among other subjects. They can also increase students' connection to nature, their enjoyment of fresh produce, and their social skills as they work together in the garden. Such gardens are easily started, though they need the commitment of multiple teachers, and they must be well integrated into the curriculum for their benefits to be fully realized.[15] Without such support, they often become a single-focus enterprise rather than a multifaceted means of educating about systems thinking.

School-based water-quality monitoring in local streams and lakes can have similar benefits. Project WET has created a wealth of resources and training for K–12 educators to engage students with water issues in their local areas. As students experience problems in one part of a system, they begin to ask questions about other parts of the system, drawing connections and seeking solutions. Some schools use volunteers or trained college interns to aid teachers in offering such experiential programs, which enhances community connections and trust. Increasingly schools also use building renovations and operations to demonstrate sustainable design and management. Their own buildings and grounds become learning laboratories, helping students understand energy flows, cycles of resource utilization and waste, sustainable purchasing, and local flora.[16]

The more students learn about the systems on which they depend, the more they want to address the issues that threaten those systems. Developing the skills of reducing waste and consumption follows naturally.[17] Although some of these skills are taught in the classroom, many are acquired through co-curricular activities and peer mentoring. If a school has developed a rich sustainability culture to match its curriculum, then new students quickly learn the kinds of skills required to fit into the culture; but this can also create the risks associated with homogenization and moral superiority/shaming described in the last chapter. If humility skills are taught alongside frugality skills, this risk diminishes. Justified by the complexity of the systems we inhabit and the diverse backgrounds of community members, humility about our own perspectives is reasonable.

As curricula shift to embrace more hands-on sustainability engagement, the role of the teacher must shift as well. The pedagogy of place- and project-based education requires teachers to relinquish some control over students, to adapt to student interests and community goals, and to learn along with students. This can be exhilarating but also exhausting. Schools that make such pedagogy routine have come together to form the Green Schools National Network and similar groups to support the enterprise of transforming K–12 education to meet the needs of a sustainable future. The National Wildlife Federation's Eco Schools program has over fifty-five hundred registered schools.

What would a deep "green" school look like? And would it cultivate resilience skills? The Common Ground School in New Haven, Connecticut, is one of the oldest environmental charter schools in the United States, and it embodies current best practices. I have known a couple of its graduates, whose stories confirm the school's excellent reputation for both academic and sustainability excellence. Located within the city limits on the edge of West Rock State Park, the charter high school has twenty acres that include an urban farm, an environmental learning center, and a sustainably designed campus. Its 225 students are selected through a lottery from sixteen towns. Some 78 percent are students of color, and 63 percent are eligible for free or reduced-price lunches. Its doors opened in 1997 with the aim of serving a whole community, not just high school students. Its mission is to be "a center for learning and leadership, inviting people across ages and identities to connect to their urban environment, build community, grow into their full potential, and contribute to a just and sustainable world."[18]

The curriculum is placed-based and interdisciplinary in structure, though it also includes a solid range of specialized courses in arts, languages, and sciences, as well as standard AP courses. The ninth-grade interdisciplinary math, science, English, and social studies core focuses on the school's socio-ecological system and the students' roles within it. The tenth-grade integrated core focuses on the larger city of New Haven, its history, its people, and its ecological impacts. Most courses have an applied, experiential component in which students practice both course-related skills and leadership. The theme of leadership permeates the culture—everyone is understood to be a leader. Each student must develop an electronic portfolio that chronicles leadership growth across four years. By the time students are seniors, they are ready to take on a major project on a social or environmental justice issue in New Haven, having learned about leveraging change in systems throughout their four years.

More than a third of the students each year are employed as part of the school's green jobs corps, building career skills while strengthening the community. Many work on the school's farm, in its environmental education programs, or on campus maintenance. The farm is thoroughly integrated into the curriculum and the larger community. It produced over ten thousand pounds of fresh produce in 2021, which supports the dining hall, a large CSA, and a mobile market. Anyone can visit the farm on Saturdays, see the sheep, pigs, and chickens and learn about sustainable agriculture practices. After-school programs both on and off campus serve over one thousand kids a year. Adding in its summer camps, field trips, family programs, and professional development workshops, the school estimates that over eighteen thousand people participate in its programs. The Common Ground School is educating a community, not just high school students. It is scaling up the impact of education.

By teaching skills associated with leadership, sustainability, and justice, the school covers most of the resilience skillful habits we have reviewed. Through learning about the social and ecological dynamics of their location, students acquire solid introductory systems skills that they use to leverage change through their projects. The full range of collaborative skills is cultivated in team projects, service requirements in the community, and the green jobs corps. Humility and frugality skills receive less direct attention, though elements of these skill sets are explicitly cultivated, including reflective skills, decentering-self skills, self-restraint skills, and self-provisioning skills. The school's strong sense of community creates a culture where resilience skills are actively reinforced. This highly applied education also yields strong academic results, including a 92 percent four-year school completion rate, with 97 percent of seniors accepted to college—both well above the Connecticut average.

The Common Ground School is small and distinctive, but elements of its model are easily replicable at larger schools. Its leaders are helping schools across Connecticut integrate school gardens into their curricula. Despite such strong models for cultivating key resilience skills in PreK–12 schools, many schools are only teaching the most basic versions of these skills or offering an occasional special program that is not integrated into the standard curriculum or the school culture. To use formal schooling to scale up emphasis on the skills that will help us flourish in these times, we need state educational standards to more clearly support their development, and we need more models that effectively integrate such skills development into the teaching of required subject areas.

One promising move in the direction of shifting educational standards is the development of the Next Generation Science Standards (NGSS), which include robust treatment of the functioning of the systems on which we depend and the fragility of these systems in our context. The standards specify numerous systems science skills for K–12. For example, one standard says that high school students should be able to "use a computer simulation to model the impact of proposed solutions to a complex real-world problem with numerous criteria and constraints on interactions within and between systems relevant to the problem."[19] The standards were released in 2013, and by 2022, twenty states had adopted the standards, and twenty-four other states had created similar versions of the standards based on the source for the NGSS, the National Research Council's *Framework for K–12 Science Education.*[20]

Teachers are often constrained by competing demands for their time, evaluations based on test results, and budget issues, so even if they see the value of resilience skills education, they may not pursue it. It can be very difficult to motivate robust resilience education in one's local schools. School systems would need to free up time and support for teachers to experiment with applied community-focused pedagogies. And they need to build school cultures that reflect the norms that will reinforce resilience skills. These last two requirements depend on having school leaders who support such educational changes. But sometimes all it takes is one visionary and energetic teacher, with some public support, to demonstrate how experiential resilience education can begin to transform children and a region.

It is unlikely that all the resilience skillful habits will be directly addressed in a curriculum, but that is not necessary to leverage the kind of large-scale cultural change that will promote broader flourishing. Collaborative skills are the most likely to be widely taught in PreK–12 settings. Systems-thinking skills are also likely to receive attention because they are cognitive skills and linked to highly salient problems. Those humility skills that are elements of critical thinking—soft conviction and reflective skills—could easily be taught across the curriculum, but here the momentum for change is relatively weak, and countervailing forces are strong, especially our cultural bias in favor of action and conviction. Frugality skills are most likely learned outside the curriculum if at all.

Fortunately, developing stronger systems thinking and collaborative skillful habits is likely to motivate lifelong learning of humility and frugality skills. As we have seen, most collaboration requires humility skills. The more we understand how complex systems function and how they can be altered, the more we see the limitations of our knowledge and the need to decrease our negative impacts on systems, often by altering consumption. It would be better to address these connections within schools, but it may be more realistic to see frugality and humility skills as addressed primarily in postsecondary education.

Higher Education

As we move to higher education, we find different challenges to emphasizing cultivation of resilience skills. Higher education is a pluralistic enterprise. It contains a wide variety of kinds of institutions, such as arts schools, mission-driven liberal arts colleges, community colleges, and research universities. This pluralism makes higher education unwieldy as a vehicle for character cultivation of any sort. Moreover, the increased cost of much higher education has made a focus on career preparation more important to students and parents. Most universities are highly fragmented, with different parts of the curriculum addressing skills that might appear to compete with resilience skills—for example, writing, critical thinking, and quantitative skills, as well as skills associated with competence in a discipline.

In the past, religiously affiliated colleges often did emphasize character education across parts of the curriculum, and a few continue that tradition today. In addition, several university programs focused on character development have been created in recent years, most notably the Jubilee Centre for Character and Virtues at the University of Birmingham in the UK. Over a decade, the Jubilee Centre has sponsored numerous research projects and policy documents, including *Character Education in Universities: A Framework for Flourishing*, produced in collaboration with the Oxford Character Project.[21] This report argues persuasively that cultivating character is a powerful way to rebuild public trust in universities, and it describes how character education might be pursued systematically at the university level. The practices it describes are similar to those used in secondary education, including explicitly exploring the skills involved in desired traits, reflecting on performance of those skills, and developing mentors and norms that reinforce the skills. Such practices could easily be included in programs that have a strong experiential component. The report maintains that character is inevitably cultivated across the university, but mostly it is unintentional, haphazard, and reflects the norms of the larger society. A school cannot avoid influencing character formation, but being intentional and systematic about the process produces better educational results.

Even if there is a strong case for postsecondary character education, faculty and administrators are unlikely to be persuaded to embrace it unless it is combined with a focus on addressing major challenges that society faces. If successfully navigating such challenges involves cultivating key skillful habits, then these skills will be a natural part of the curriculum and co-curriculum. As we have seen, the challenges of developing a more sustainable and just society involve a range of resilience skillful habits, and indeed higher education is increasingly focused on these challenges. Student concern about sustainability and social justice is very high, which is pushing the development of curricula in promising directions.

Anthony Cortese, founder of Second Nature and the American College and University Presidents Climate Commitment, powerfully articulated the rationale for infusing sustainability throughout higher education curricula:

> Higher education institutions bear a profound moral responsibility to increase the awareness, knowledge, skills, and values needed to create a just and sustainable future. . . . [Higher education] prepares most of the professionals who develop, lead, manage, teach, work in, and influence society's institutions, including the most basic foundation of K–12

education. Higher education has unique academic freedom and the critical mass and diversity of skills to develop new ideas, to comment on society and its challenges, and to engage in bold experimentation in sustainable living.[22]

Second Nature focuses primarily on increasing leadership support for sustainability education, but it helped to create the Association for the Advancement of Sustainability in Higher Education (AASHE) in 2005, which provides training and networking for faculty, staff, and students. Within a decade, AASHE grew its membership to include more than 20 percent of the approximately four thousand degree-granting higher education institutions in the United States. AASHE introduced the Sustainability Tracking and Rating System (STARS) that sets standards for sustainable practices throughout an institution. These standards—the higher education equivalent of LEED building standards—have guided institutional transformation. STARS ratings and sustainability rankings such as Princeton Review's *Guide to 361 Green Colleges* and *Sierra* magazine's *Cool Schools* issue have created friendly competition between universities and helped students find schools that are in the sustainability vanguard. According to a 2021 Princeton Review survey of eleven thousand prospective students, 78 percent indicated that "having information about a college's commitment to the environment would affect their decision to apply to or attend a school."[23]

The sustainability and social justice movements in colleges and universities have taken off in the last two decades. Students deserve a lot of the credit for this, since they have pushed hard for concrete achievements, not just verbal support. Sustainability and social justice have become important themes in a wide range of programs, from business and agriculture to chemistry and community development. Such curricula may not sufficiently emphasize development of resilience skillful habits, but they acknowledge that changing habits of thought, feeling, and behavior are central to addressing sustainability and justice problems.

The resulting cross-disciplinary curricula are forms of character education without the name. Skills of systems thinking are becoming much more widely taught. Collaborative skills are receiving more nuanced treatment, as their demand in many professions grows. Some humility skills, like recognition of potential biases and blind spots, are increasingly taught as part of social justice and critical-thinking curricula, though they may not be broadly reinforced across the curriculum. Most schools do not yet embrace an explicit framework that reinforces resilience skillful habits and connects them to flourishing, but this next step is easy to take. It may be harder to incentivize faculty and staff to role-model these skills.

How could we more fully embody the vision for education that would promote a just and sustainable world that Cortese champions? A few years after I arrived at Green Mountain College in 1996, that was the kind of question I was asking. The college was poised for such change; it already had a mission focused on the environment, which soon evolved to focus on sustainability and justice. There were few road maps then, but our faculty and staff were inspired by the task of living the mission and willing to risk promising experiments that might guide our answers. We pursued a kind of adaptive management within a college setting. I have summarized elements of that odyssey elsewhere.[24] Here I will describe three initiatives that cultivated resilience skills—monster courses, sustainability skills courses, and the "Delicate Balance" capstone. These illustrate the kinds of changes we need to make, not what will work in all institutions. For more ideas about sustainability education in general see Mitch Thomashow's excellent book, *The Nine Elements of a Sustainable Campus*, which chronicles the development of answers he championed as president of Unity College.[25]

Monster courses were team-taught twelve-to-fifteen-credit offerings that immersed students in a set of systems issues in a bioregion. Since they involved most of the students' time during a semester, they could involve multiday field trips engaging with regional stakeholders and seeing firsthand the impact of issues. For example, in Envisioning a Sustainable and Resilient North Country (2016), seventeen students and four faculty spent sixteen weeks trying to grasp the environmental, social, and economic dynamics affecting lives in Vermont and upstate New York and to crystallize elements of a twenty-five-year plan aimed to move the region toward greater sustainability. In talking to over fifty stakeholders and seeing and hearing their perspectives and stories, students learned how to listen and integrate diverse takes on regional problems and solutions. They learned how to facilitate a planning process and practiced the roles of facilitator and participant while crafting a draft plan that they presented at the Vermont statehouse. With the assistance of Douglas Gayeton at the Lexicon Project, they learned how to create information artworks to communicate sustainability concepts in emotionally compelling human terms. Their twenty-four artworks traveled around the state and to a national conference. And they learned about Marshall Rosenberg's nonviolent communication, using these skills to engage with participants in an emotional conflict over resettling Syrian refugees in Rutland, Vermont, and to develop a website that fairly represented the conflicting views.

No short description of a monster course can capture the closeness of the community that develops throughout the course, the conflicts that need to be navigated, the opportunities for deep mentoring that faculty have. The public nature of the student work and its importance to others in the region motivate students to push themselves beyond normal expectations, often to the point of exhaustion. It also requires a lot more from faculty than an ordinary course, though it is inevitably a tremendous faculty development experience, worth all the effort and angst of teaching while learning alongside the students. These courses seem to have greater long-term impacts on student skills development and on instructors' interdisciplinary systems knowledge than any other curricular offering I know. The planning and logistics can be time-consuming, but the concrete engagement with issues that matter and with students as whole persons, not just minds in a classroom, more than compensate for most faculty. Such courses provide tremendous opportunities for cultivating a wide range of resilience skills for both students and faculty.

While monster courses emphasized systems, collaborative, and humility skills, one-credit sustainability skills courses cultivated frugality skills in the self-provisioning, repair, reuse category. These courses aimed to provide introductory-level proficiency in a suite of skills, such as solar panel installation, home energy audits, foraging for wild foods, food preservation, bike repair and maintenance, and meditation. The college offered over fifty different skills courses, usually taught by local practitioners. They were wildly popular because students developed hands-on skills that they could use immediately. We had to limit the number of these courses that students could count toward graduation, but we also created a certificate that students could earn by passing six skills courses. Initially I thought of these courses as the sustainability equivalent of the old physical education courses. But eventually I came to see that students' hunger for concrete nonacademic skills ought to be honored as an integral part of sustainability education. Learning such skills sometimes led to first jobs or summer work. A local solar panel installer taught the course regularly, so he could hire his best students.

The capstone general education course, A Delicate Balance, was taught by multiple faculty every term. I taught it twice a year for five years after I stepped down as provost. It had three main elements. First, students read a memoir and wrote their own reflective essays on who they are, what they had learned in college, their strengths and weaknesses, and how they envision their lives unfolding. In discussing memoirs like Ta-Nehisi Coates's *The Beautiful Struggle*, students came to see how we can reflectively craft a story of our lives and why that reflective process might be important. A surprising number of students had not thought much about why habitual reflection is valuable. Though many felt they failed to find a "balance" between humility and pride when reflecting on strengths and weaknesses, most were proud of the stories they crafted and shared them with family and friends.

Second, throughout the course, students planned and executed an applied project (often in groups) that enhanced some community, using their skills to make a difference in others' lives. They read about project management, consulted with stakeholders, developed proposals and budgets that had to be revised and approved, and shifted project goals in light of constraints. Although many chose to organize one-shot events for the campus or town, some created lasting contributions to the college—for example, a stunning outdoor classroom, an ongoing spring break community service project, or an important policy change.

The course's third element aimed at enhancing empathy and understanding of those who might resist well-meaning efforts to make a community better. It was a sustained discussion of the wedge issues around views of freedom, justice, authority, and diversity that divide many communities and make it difficult to bring people together to create positive change. It emphasized the rationales people had for their differences and encouraged students to confront ways in which their own positions might bias them against others in the community. The theme of balance ran through the course, and we interpreted that theme as applying to how we might navigate some of the tensions between the social norms of different subcultures and bring into harmony competing elements of our characters. The three elements of this course interacted to create at least one place in the curriculum where discussion of character could be linked to specific skills students had learned throughout their education and where they had to think about what flourishing meant to them.

Almost all students were aware that Green Mountain College was a kind of bubble with its own set of social norms that were very different from those of the surrounding culture. Some did not want to leave the close-knit community it created, but others felt stifled by the implicit pressure to conform. Although students appreciated many forms of diversity and inclusion, our surveys revealed that political diversity was not among them. The benefits of strong norms were evident in the way they reinforced the skillful habits we thought important, but the downside of the resulting tendency toward homogenization was disturbing. Some of us worked hard to powerfully present unpopular perspectives. Eventually, though, I concluded that we had failed to cultivate the soft-conviction skills that are a crucial element of humility in our larger cultural context.

Scaling up approaches to higher education that would successfully cultivate resilience skills would require major changes in most schools. Parts of the general education curriculum would need to have an explicit normative core, which is reinforced by the co-curriculum and echoed in many majors. Experiential and project-based education would be much more widespread. Faculty would engage as whole persons with students, rather than primarily as subject matter experts. They would be role models in more than their disciplinary expertise. Faculty culture would become less adversarial, more humble, and more collaborative. Faculty and staff would become comfortable reasoning about which character traits and norms to emphasize in a context. And presidents, provosts, and deans would set a cultural tone through word and deed that reflects the skillful habits students are learning. All of this is compatible with honoring important forms of pluralism on campuses if the kind of binocular vision I have defended is part of the package and humility is emphasized.

Even if an increasing number of schools continue to move in the direction indicated above, the percentage of graduates who acquire nuanced resilience skills will be not be large, and there will be a significant time lag before they become leaders who can influence the direction of other institutions. The speed with which we must shift common behaviors toward greater sustainability and justice requires encouraging adults to strengthen skillful habits outside of formal educational settings. Here we turn to a much broader conception of what education involves.

Building Skills out of School

Much of our learning occurs outside of formal educational contexts. We learn by attending programs, working together to achieve some goal, doing our own research, and talking to our neighbors. We can think of such opportunities as being on a continuum from more- to less-structured educational experiences. The more structured they are, the more the content is controlled and results are predictable. But the more a program is structured to meet explicit learning goals, the less likely it is to reach beyond those who already want the skills it teaches. Since the scaling up we seek requires education to reach beyond those who seek it, the less-structured forms have important value even though they involve less quality control.

As the rate of social and technological change increases, and career changes become more frequent, we need to increase our capacity to learn new skills across our life span.[26] A key part of the UN's Decade for Education for Sustainable Development was a renewed emphasis on lifelong learning and the creation of a host of nonformal sustainability-related learning opportunities. The internet has made the dissemination of such opportunities much easier. We are rapidly approaching the point where we have a wealth of opportunities for learning new skills, and the primary issues we face are lack of time and inclination to make the most of what is available.

On the more-structured end of the nonformal education continuum, we find many short courses aiming to teach resilience topics. For example, we could learn more about systems thinking, effective collaboration, personal resilience skills, the circular economy, or sustainable development by taking a free massive open online course (MOOC).[27] In most urban areas, one can also find relevant skills trainings at local community colleges and many other organizations. Where these are applicable to resilience skillful habits, courses will tend to focus on specific skills, not on the global traits supported by such skills.

Such programs can also provide strong theoretical understanding of the foundations of resilience skills, especially those in demand by organizations, but often their impact is short-term and merely intellectual. The deeper learning required for skills development and behavior change tends to be directly proportional to the time and effort required by the learner, the amount of feedback one receives on performance, and the application of learning in a place. Another limitation is that general information provided through courses rarely engages learners in mastering the application of skills in a context—for instance, applying systems thinking to our own communities. These are inevitably first steps toward more experiential learning.

For more place-focused short educational programs, people often look to nonprofit organizations that offer experiential learning in their regions. For example, families may attend programs at nature centers to learn about the local ecosystem. These kinds of programs may teach a few specific skills, but they are at best a gateway to richer skills development. It is much rarer for families to engage with programs about local social or economic systems, in part because fewer of these exist. But we need to learn about adaptive governance, about how to build and repair trust, and how to address social injustice in our communities, as much as we need to learn about our ecosystems. Volunteering for nonprofits to achieve specific community goals is much more likely to cultivate collaborative, systems, and frugality skills.

We often think that less-structured learning is less effective, but one of the most powerful forms of resilience skills education occurs when citizens participate together in community planning processes and local association governance where strong facilitators guide groups through decision-making and model the skills necessary to meet the needs of diverse groups.[28] In these contexts, part of what is learned comes from community members who share their perspectives on local system dynamics, on threats to what they value, and on desirable transformations. Usually, expert knowledge needs to inform such processes, but citizens have their own expertise about where they live and what skills have successfully led to beneficial change in the past. Collaborative and humility skills are also fostered through such processes. Robert Putnam has argued that participation in local associations strengthens skills that build social capital.[29] Ideally, facilitators set norms, and participants practice listening well, managing conflict, overcoming obstacles, and helping groups move to closure and action. As we saw in chapter 3, decline in participation in local associations has been correlated with decline of interpersonal trust and the skills that support it.

Well-facilitated participatory planning processes have a wide range of benefits linked to community resilience, beyond producing a plan and generating social learning. They can build new relationships and networks that cushion shocks to a community. They also help preserve intangible qualities of a community. Refocusing attention on relationships, beauty, and environmental quality contributes to a shift away from material consumption. They can also strengthen the sense of identity of a community. Indeed, Bryan Norton argues that we can only know what sustainability is in a community after we have entered into collaborative processes to decide what community attributes must be preserved for future generations.[30] I have found that these processes have done much to strengthen hope and a sense of shared humanity.

Participatory processes can occur at multiple scales. Some long-term, large-scale processes like the Chesapeake Bay restoration program have had a dramatic impact on a region through their educational efforts; but smaller-scale processes often engage a broader range of participants. The hundreds of small community energy committees in Vermont towns are transforming citizens' thinking about our energy future. Watershed groups in California and elsewhere have had great success educating a wide range of stakeholders while restoring salmon to rivers like the Klamath.[31] This kind of collaborative decision-making often generates local learning at its best. Of course, participatory processes can go badly. They can be hijacked by powerful interests, they can be frustratingly slow, and they can result in compromises that few support. Much depends on the quality of facilitation and the skills of participants.

Opportunities at Work and in the Sanctuary

Businesses and religious organizations offer a wide range of nonformal educational opportunities, such as job training and spiritual practices workshops. They also often have strong subcultures that encourage internalization of the skillful habits they value. Both have the potential for building resilience skills.

On average, adults in the United States in 2020 spent 1,767 hours working over the course of the year.[32] The organizations where we work often shape who we are. Most work environments have cultures to which we must adapt. Some organizational cultures may be dull and deadening, others toxic, but many engage us and create opportunities for deep growth. Large organizations usually deliberately craft cultures that motivate employees to be highly productive. Although businesses frequently foster competition, conviction, and consumption habits, they usually need effective collaboration within the organization and with multiple stakeholders outside it. They certainly need employees who have a solid understanding of the socio-ecological systems in which their business operate. As a result, many businesses build cultures that emphasize elements of the resilience skillful habits. The sustainability tsunami that is rapidly changing business practices plays a key role in motivating shifts toward cultures that reinforce resilience skills.

Some years ago, I presented my work on resilience skills to a group of Fortune 500 company sustainability officers. They resonated with the importance of the skills but feared that the reality of their corporate cultures would not support such skills development. With a colleague, Matt Mayberry, I honed the argument that a business culture that cultivates resilience skills would give a company an edge as we move through the sustainability transition. We looked for a company that had experimented with creating such a culture and found one in Green Mountain Power (GMP), the largest utility in Vermont. The CEO, Mary Powell, had arrived at a company that was imbued with a very traditional utility culture—very hierarchical, slow, internally competitive, and definitely not frugal. When we interviewed her, we were surprised to learn that she thought *all* of the resilience skillful habits were critical to the success of her business and that much of what she had accomplished came from deliberate attempts to shift the culture in their direction. These skills did not just make GMP a more sustainable company; they made it a more effective company, with better relations with customers and regulators and more innovative and frugal practices.[33]

But how can a workplace shift its cultural norms and employee skills to better align them with the challenges it faces and its organizational goals? Unlike schools, where students can practice and make mistakes, risking only their grades, workers may risk lost promotions or even dismissal for failures, which creates significant anxiety around experimenting outside the received culture.[34] Standard business training does little to alleviate such anxiety; it typically involves employees absorbing information and then applying it in a project for their employer. Learning occurs, but the true realization of skills learned takes place afterward in a higher-risk environment.

Some effective business training uses simulations presented online over extended periods in which employees can work collaboratively, testing their understanding of new skills, seeing the results of decisions, and then applying skills in a new context. Creating a low-stakes practice field for trying out new skills speeds skill formation and decreases anxiety about change. The ten-week program Leading the Sustainability Transition (LST) uses a simulation in which teams of participants practice new skills to develop strategies for a multinational bioproducts company to reduce waste, strengthen stakeholder relations, introduce new products, and streamline operations over a twenty-year period.

Trainees learn how to leverage change in the systems in which the company oper-
ates, build more effective collaborative relations, and achieve frugal solutions;
these are crucial to enabling the company to flourish. Mistakes are inevitable and
part of the learning process. Some participants are looking for a career change
that promises to better connect their values with their jobs, but most are sent by
companies that want to green their operations.

Programs like LST can significantly strengthen resilience skills. They are rare,
but they illustrate a kind of high-quality nonformal education that could easily
be scaled up and that can be offered outside of a business context. Simulations
are not the only way to encourage practice over time with feedback; most lessons in music, visual arts, and sports have such structures. Unfortunately, few
opportunities exist for continuing lessons in resilience skills, except where these
are related to professions. For example, certificate programs that address key collaborative skills like group facilitation, mediation, and conflict resolution are easy
to find, though they tend to be designed primarily for professionals.

Historically, religious institutions have played an outsize role in cultivating
skillful habits connected to morality. Typically, they emphasize dispositions to
act morally rather than skills development, and their methods are hortatory.
Text, sermon, and song aim to inspire practices. Such methods do encourage
skills development, though, because as believers develop habits, they seek to
improve outcomes by refining the skills involved in applying the habits. I may
learn first that I ought to be hopeful, but after developing the habit, I begin to
adjust my judgment about when hope is warranted and how to communicate
hope effectively.

Take, for example, the Christian tradition, which significantly influenced US
culture. For many generations the primary source of humility skills was the church.
The decentering of the self was prominent in worship and care for community.
Hubris was undermined by comparison with God's knowledge and power and by
recognition that achievements required grace. Reflection skills were given weekly
practice through sermons and adult Sunday school. Of course, a tradition of hard
conviction is associated with many denominations. Often, traditions emphasize
some elements of a complex skillful habit while de-emphasizing others. Empathy,
hope, and frugality are also elements of standard Christian virtues. Role models
for these virtues are described in classic stories, and proverbs criticize those who
lack them.

Most religions contain important strands that support each of the resilience
skillful habits, since these are strongly connected with traditional moralities.
Most also have powerful mechanisms for transmitting their cultural norms and
providing advice for those seeking the wisdom such skills enable. They are well
positioned to play a significant role in scaling up resilience skills—though in a

period of religious decline, many faith groups have chosen not to challenge our dominant skillful habits to avoid losing members.

Independent Learning and Sharing

At the least-structured end of the education continuum, we find that we are both educators and learners. We access information wherever we can find it, whether in books, online, or from friends. We try new ventures and relearn skills our grandparents took for granted. And along the way, we build community by sharing what we know. I was never taught how to grow vegetables, but with much reading and many failures, I have learned enough to preserve a vegetable harvest big enough to carry me through winter. We acquire most of our self-defining skillful habits largely through informal education. The vast majority of education occurs in hyper-local interactions. Each of us has a critical role in the community education process.

The renowned climate scientist Katharine Hayhoe argues that the most important thing each of us can do to address climate change is to talk about it. In our action-oriented culture, this seems counterintuitive; we worry about "all talk, no action" critiques. But Hayhoe wisely notes that to make the changes necessary for climate solutions, we need a social movement, and that talking with others is the first step in movement building. She does not mean we should lecture others about climate facts and solution prescriptions; indeed, the evidence suggests that such lectures rarely produce the change we want.

Her recipe for effective action is easy, humble, and effective. We should listen to others and learn what they love. We then connect the dots between their passions and the negative impacts of climate change. We avoid the hubristic implication that they become more like us, and instead start with who they are. Once we have built the bridge to climate issues, then we can empower our interlocutors by describing solutions that are easy for them to take. In such a conversation, we are educators but not "teachers." We are fellow travelers, sharing parts of our journey with others with whom we have created an initial bond.

Such conversations take time, and often initially they have limited impact. Hayhoe is not providing a silver bullet for movement building, but her excellent book, *Saving Us: A Climate Scientist's Case for Hope and Healing in a Divided World*, provides many effective strategies for breaking through polarization and engaging the disengaged.[35] She also provides invaluable lessons about how to be more effective informal educators, lessons that reinforce our resilience skills. Empathize with where others are, build trust, and foster hope. Be humble, even when expressing our convictions. Use our understanding of systems to show how climate change affects what others care about. And where appropriate share our frugality practices.

With the opportunity to serve as educators comes significant responsibility. The "quality control" in informal conversation can be very low. Part of our task must be to raise the standards for knowledge in our discourse, emphasizing the importance of data, scientific consensus, and careful systems thinking. Unfortunately, half-truths and misinformation shared on social media seem to travel much faster and farther than more sober fare. Furthermore, our conversations must bridge divides and build a shared reservoir of understanding in a community. If we only share with those like us, we may reinforce polarization and unwittingly contribute further to marginalizing others. Communication style, emotional tone, and overall intention matter a great deal. Adults generally do not like to be taught without their consent, and no one likes to be manipulated or looked down upon. This learning and sharing must be multidirectional, with listening being as important as speaking.

I am a teacher by training, and I love the challenge of designing a powerful learning experience for students. But I find the serendipity of the informal education process at least as rewarding. The neighbor who stops by and asks whether I like my solar panels gives me an opportunity to talk about their benefits and the short payback period. I learn from her about her composting system. As each of us shares our passions and practices, we also refine our skills. Our conversations at church suppers or local basketball games can influence people who would never read a book on resilience skills and would reject sustainability as mere liberal politics. The reach of our informal education can be as broad as the formal K–12 system, and its cumulative impact can change culture at greater speed.

The Prospects for Scaling Up

Momentum is building for scaling up cultivation of some resilience skillful habits through education, but we are still far away from social norm change. If you are reading this book, you know that forces of inertia and short-term self-interest remain very powerful. Often movements flag when our attention turns to other issues or when well-intentioned mistakes derail promising strategies. Even if a movement grows, its goals may recede, or it may be co-opted by powerful groups with other aims. Our cultural emphasis on decisive action and short-term publicizable results makes any intentional long-term transformation of cultural norms difficult. We need a mix of concerted effort and good luck to achieve the shift in cultural norms that will reinforce enough acquisition of resilience skills.

Admittedly the tone of our times seems to be pushing us in the wrong direction—toward more competition, more conviction, and more material consumption. But I have to remind myself that mainstream media and social media tend to highlight the shrill fringes of our social fabric. My experience of people from across the political spectrum strongly suggests that a majority would welcome shifts toward norms that reinforce resilience skillful habits. They are influenced by traditions that support frugality, humility, and collaboration and are aware that the challenges we face require the skillful expression of these virtues. They would also support the kind of binocular vision that integrates them with contrasting virtues.

The magnitude of the changes necessary in formal education is great, and the potential for problematic implementation is significant. Just as the goal of limiting global warming to 1.5°C is now a long shot, so is the goal of scaling up cultivation of resilience traits in a short timeframe. But we are not betting on long shots, as if we stood on the sidelines. We are investing in very important goals with the understanding that we do not need total transformation for the effort to be worthwhile. Every small increment of progress makes a difference. Good education for resilience skills turns out to produce high-quality learning in the full range of standard academic subjects. We need not trade off academic quality for resilience excellence. If the case for cultivating resilience skills to promote flourishing is not convincing for some, then the case for its promoting academic quality should fill some of the gap.

Each of us can contribute to the educational transformation. Wherever we are in our life spans, we can pursue our own cultivation of resilience skills, building demand for more learning opportunities and strengthening momentum in the movement. We can talk to our friends and neighbors about what we are doing, as Hayhoe urged. We can model resilience skills with appropriate humility, avoiding the pitfalls of implicit superiority and shaming. We can volunteer in schools, helping teachers to implement the applied project-based education that enables students to practice skills, and we can involve our organizations in such projects. We can actively engage with school boards, and if we have kids in school, we can participate in parent-teacher organizations, supporting educational changes that align with quality character development in schools. And lastly, we can express gratitude to the teachers who helped strengthen the skills we value.

EPILOGUE

One of the great ironies of our age is that while our technological powers are tremendous, our ability to address collective-action problems is small and declining. Solutions to our critical challenges often seem beyond our reach; yet we cannot let despair rob us of the motivation needed to double our efforts when the odds of success diminish. How should we think productively about how to navigate a future that seems both highly uncertain and gloomy?

In some ways, we are well situated to think rigorously about this question. We know a fair amount about the complex causal linkages between ecological systems, regional economies, and governance. We have a wealth of historical knowledge about what has happened in societies that were approaching the release phases of their adaptive cycles. Psychology has provided significant insight into the variables that affect flourishing, and we have access to the wisdom of many traditions about how to live well. And yet, the number of options for promoting our welfare is huge in this age of information overload, making the decision landscape muddier. The scale and complexity of our global challenges, our political issues, and our moral obligations cast us into a sea of competing considerations. The welter of specialized knowledge and competing analyses leads to greater uncertainty. To compensate, we must creatively lump potential futures into a few likely scenarios in order to simplify decision-making.

Scenario planning is widely used by governments, businesses, and militaries to identify the most promising strategies to adopt. Climate scientists use a range of scenarios to tame uncertainty by highlighting different pathways we might take and show how these are likely to affect global warming. We can adapt these climate scenarios to help us think critically about whether to invest our time in strengthening resilience skillful habits. Brian C. O'Neil and colleagues have developed five scenarios for this century, called Shared Socioeconomic Pathways (SSP), based on our current understanding of the linkages between many variables that shape the functioning of our global socio-ecological systems.[1] These scenarios have been through a lengthy peer review process. The brief narratives that characterize the scenarios sketch the results of complex feedback loops connecting governance, resource usage, population, economic growth, inequality, and many other aspects of societies. Although these were created to reflect how different global pathways will affect climate change mitigation and adaptation, they nicely serve our purpose of seeing how options for flourishing can play out.

I propose we focus on two SSPs that occupy opposite corners of the scenario space with respect to breadth of flourishing. Scenario SSP1, "the green road," outlines an optimistic future in which societies approach just sustainability, and SSP3, "the rocky road," describes a pessimistic picture in which societies descend into international and internal conflict that sidelines sustainability. The other three—high inequality; fossil-fuel-driven development; and the middle road— have importantly different implications for climate mitigation and adaptation, but less significant implications for our approaches to flourishing. How would a decision to invest in cultivating skillful resilience habits likely affect flourishing in each scenario?

In the green road scenario, strong investment in environmentally friendly technologies like renewable energy gradually reduces negative environmental impacts. Societies increasingly internalize the costs of environmental and social harms, further reducing these. In addition, human well-being rather than economic growth is increasingly the focus of development; consumption shifts toward less energy- and resource-intensive goods. Population growth decreases with greater investment in health and education. Inequality also decreases both internationally and within countries. Lastly, "Management of the global commons slowly improves, facilitated by increasingly effective and persistent cooperation and collaboration of local, national, and international organizations and institutions, the private sector, and civil society."[2] In short, we move toward sustainability with reasonable speed.

To take this road, global leaders would need to become much better at collaborating well over time. They would need the support within their own countries for sacrificing some perceived short-term national interests in order to prioritize goals that can be supported by the majority of countries. In democratic countries this can only happen if populations understand enough about system dynamics at a global scale, the tradeoffs necessary for collaboration at that scale, and the value for all of a more just distribution of the benefits and costs of progress.

Because humility is necessary both for effective collaboration and for recognizing that our systems understanding is always partial and flawed, our leaders would also need strong humility skills, which must be supported by the populace. Similar points apply to frugality norms. The shift away from high material consumption toward social, aesthetic, and spiritual goods, and the emphasis on decreasing inequality and increasing overall well-being, dovetail with the strengthening of key frugality skills. Given the level of popular support necessary for leaders to forge far-reaching sustainability agreements, their cultures are likely to be characterized by norms that reinforce resilience skillful habits.

In this context, we would be more likely to flourish as a result of cultivating resilience skillful habits. Doing so would support a global movement toward sustainability, a movement that enhances flourishing for other humans *and* for populations of threatened nonhumans. This is likely to provide meaningful engagement, participation in something larger than self, and a sense of achievement. We would be developing skills that are valued by most of our communities and that strengthen our relationships, both of which contribute to flourishing. Our paths through life would be eased as we work on skills that are rewarded by our institutions. In an important sense, we would be swimming with the current. Of course, there would also be some costs. We might feel too confined by norms that emphasize collaboration and humility. I might miss activating my inner rebel that loves pushing against the current. I might also miss some of the conveniences that enable more time thinking and writing. But these costs seem to pale in comparison with the gains.

Others who are invested in the currently dominant skillful habits would no doubt pay more dearly for the shift in cultural norms, but individualistic, competitive, and conviction skills would still be valued. The green road is not a scenario in which capitalism is abandoned or competitive sports downplayed. Nor is it all peace and love, a hippie paradise. It will contain plenty of competition and conflict. It will require the binocular vision that has been a theme throughout this book, and the wisdom to know when to engage our competitive skills and when to invest heavily rather than frugally. Our inevitable mistakes will strengthen our humility skills, but our soft convictions must be strong enough that we can hold our course, when deviating would garner more immediate accolades.

But what about the other scenario, the rocky road strewn with strife? In this scenario global institutions are weak, and cooperation between countries limited to what serves short-term national agendas. Conflict between and within countries becomes more frequent, and security becomes more important. Economies become more regulated as countries try to provide their own energy and food; technological development diminishes and is not widely shared. Inequality remains high both between and within countries, and low concern about environmental issues strongly affects poorer populations. Most people also struggle with lack of investment in community health and education. Human population grows in poorer regions that have limited access to family planning. Consumption remains energy-intensive and materially focused. Fossil fuel use remains high in countries where this is the most accessible form of energy. Authoritarian governance spreads as a result of increased conflict and resources shortages. Social unrest is common, and progress toward sustainability is slow. Fires, flooding, and extreme heat are much more common, but only rich nations can afford to invest heavily in adaptation. The widespread suffering of multitudes will be highly salient. This future is deeply disturbing, but it is not total collapse, at least in the developed world. It is, however, a world in which serious release occurs frequently·on multiple scales.

In this scenario, the norms reinforcing our dominant skillful habits do not change in most places. Competition will become more prevalent, and the stakes associated with winning and losing will seem higher. A me-first form of individualism may become even more common, and leadership full of conviction will seem a bulwark against the surrounding chaos. Many more will practice frugality out of necessity, but they are likely to crave the abundance still found among the elites.

Those who have worked to develop resilience skills are likely to gravitate toward subgroups with similar orientations—smaller communities removed from the sources of conflict or organizations with cultures that counter the dominant trends. Their systems skills will enable them to understand the impacts of larger-scale systems and help them to craft resilient communities within that context. Their collaborative skills will enable them to grasp the opportunities to build relationships with diverse groups and avoid intractable conflict. Their capacity for hope will carry them through the chaos with less psychological scarring. Their humility skills help reduce the tendency for conflict absorbed from the larger culture and increase the gratitude for the many goods they still find in life. Effective reflection will tend to dampen the negative impact of large-scale events and heighten appreciation for the strengths and weaknesses exhibited in the all-too-human efforts to manage life on the rocky road.

Their frugality skills will enable them to find plenitude within the resource constraints that will characterize the lives of most people. They will be able to happily make do with what they have available. But perhaps most importantly, they will be focused primarily on intangible goods—strong relationships, natural beauty, and their own character development—which can still be savored no matter what else is happening. It is easy to get caught up in the dramas of the day, but we can cultivate the skills of stepping back and appreciating all that we have. To be sure, large parts of the natural world will be severely degraded in this scenario, but so much beauty will remain. As Mary Oliver reminds us, "Sometimes I need only to stand wherever I am to be blessed."[3] Even in a large city, bits of nature can be a balm. Warm words from a good friend can soothe our uncertainties. The touch of a lover can transport us farther than the fastest plane. More than almost anyone, Mary Oliver has shown us how to find the gifts we are bequeathed even in a broken world.

But would these skills give us enough to really flourish along the rocky road? As I have repeatedly said, they are no silver bullets. Yet we would be much more likely to flourish than people whose character reflects the norms reinforcing individualistic competition for increasingly scarce material abundance. On average, such people will be less likely to effectively navigate the tumult, to craft supportive communities, and to find the opportunities for meaningful engagement provided by our challenges. As I have argued in previous chapters, our skills would likely be more than enough to support genuine flourishing even amid the diminishing prospects for peace, justice, and sustainability.

In both these scenarios and those that fall between them, we are likely to be better off if we cultivate resilience traits. Such skillful habits are better aligned with the challenges and opportunities that we find in the age of climate change. In a different age of rapid growth and plentiful resources, our culture's dominant skillful habits might better align with the times. In all parts of the adaptive cycle, it is wise to develop contrasting sets of skills, though that is hard, given our limited time and capacity. Even if we pursue the kind of binocular vision I have advocated, we will inevitably have to choose where to concentrate our character development. Being in tune with the times will help us to flourish.

The resilience skills have been important in many traditions that inform the United States and other cultures. We are not trying to invent something new, but rather to retrieve and reemphasize skills that were once more common. While resilience skills are in tension with our dominant skillful habits, the two sets of skills can be fruitfully combined as we develop better binocular vision that integrates them into the more robust and flexible toolboxes we need to address our challenges. A powerful global movement toward justice and sustainability provides hope that we can achieve that combination of skills.

What I most need now are guides that can help me to stay the course, to practice and occasionally fail to exemplify the character traits that will help me to flourish while we engage in the great struggle to change the structures of large-scale systems that shape our lives. Mary Oliver has been one such guide, helping me to see the gifts that surround me. Another is the philosopher of struggle, Friedrich Nietzsche. He shows us how to find joy in what is difficult, how to welcome our challenges rather than resent them. He continually employs the image of the dancer who embodies the life-affirming joy of movement amid tumult. He says, "Smooth ice is paradise for those who dance with expertise."[4] Smooth ice is dangerous for most of us to walk upon, but for some, it can bring joy. A good measure of the resilience skillful habits will give us much of the expertise needed to dance amid the dangers of the age of climate change.

Notes

INTRODUCTION

1. Release is a phase that complex systems go through after periods of growth and conservation of resources, in which a major threshold is crossed, structure is lost, and conserved energy is released. In severe cases, it is when a system collapses. I argue that we should see our society as in the late stage of the conservation phase moving toward release.

2. Martin E. P. Seligman, *Flourish: A Visionary New Understanding of Happiness and Well-Being* (New York: Atria Books, 2012).

3. Valerie Tiberius and Alexandra Plakias, "Well-Being," in *The Moral Psychology Handbook*, ed. John Michael Doris (Oxford: Oxford University Press, 2010); Ed Diener, "Subjective Well-Being: The Science of Happiness and a Proposal for a National Index," *American Psychologist* 55 (2000): 34.

4. Aristotle, *Aristotle's Nicomachean Ethics*, trans. Robert C. Bartlett and Susan D. Collins, repr. ed. (Chicago: University of Chicago Press, 2012).

5. See for example Amartya Sen, "Capability and Well-Being," in *The Quality of Life*, ed. Martha Nussbaum and Amartya Sen (Oxford: Clarendon, 1993), and Martha C. Nussbaum, *Creating Capabilities: The Human Development Approach* (Cambridge, MA: Belknap Press of Harvard University Press, 2013). Well-being is a close cousin to flourishing and is usually explained in the same way.

6. This kind of view of flourishing or well-being is typically called an objective list theory. For a rigorous defense of this view see Christopher M. Rice, "Defending the Objective List Theory of Well Being," *Ratio* 26 (2013), https://doi.org/10.1111/rati.12007.

7. Current measures of well-being include questions relating to financial security and health. These are one way of assessing whether someone has basic needs met. See Dorota Weziak-Bialowolska et al., "Psychometric Properties of Flourishing Scales from a Comprehensive Well-Being Assessment," *Frontiers in Psychology* 12 (2021), https://www.frontiersin.org/articles/10.3389/fpsyg.2021.652209.

8. Arne Naess, "The Shallow and the Deep, Long-Range Ecology Movement. A Summary," *Inquiry: An Interdisciplinary Journal of Philosophy* 16 (1973), https://doi.org/10.1080/00201747308601682. For a solid overview of deep ecology see Bill Devall and George Sessions, *Deep Ecology: Living as If Nature Mattered* (Salt Lake City: Gibbs Smith, 2007).

9. Christopher Peterson and Martin Seligman, *Character Strengths and Virtues: A Handbook and Classification* (New York: American Psychological Association and Oxford University Press, 2004).

10. Situationists have argued that stable global character traits rarely exist and do not explain behavior; see for example John M. Doris, "Persons, Situations, and Virtue Ethics," *Noûs* 32 (1998). I have been convinced by numerous responses that these arguments are not telling; see for example Eranda Jayawickreme et al., "Virtuous States and Virtuous Traits: How the Empirical Evidence regarding the Existence of Broad Traits Saves Virtue Ethics from the Situationist Critique," *Theory and Research in Education* 12 (2014), https://doi.org/10.1177/1477878514545206.

11. Philosophers will recognize this book fits squarely in the virtue ethics tradition, which has had a tremendous resurgence in the last thirty years. I am deeply indebted

to many authors writing on environmental virtue ethics, but I will rarely use the term "virtue," and I will avoid discussion of philosophical issues associated with virtue theory. I intend the phrase "beneficial skillful habit" to be the rough equivalent of "virtue," but I hope it will be less daunting to a general audience. Virtue ethicists may think that by highlighting skills I am placing insufficient emphasis on the inclination to use the skills in ways that are good. I agree that this inclination is important, but one of the best ways to foster such inclinations is to cultivate enough relevant skills that their successful application increases the inclination to use the skills. Directly trying to change people's inclinations often seems like manipulation and can foster backlash. I will address this point in more depth in the last two chapters of the book.

12. In chapter 1 I will provide a more extended treatment of resilience.

13. Kathryn M. Connor and Jonathan R. T. Davidson, "Development of a New Resilience Scale: The Connor-Davidson Resilience Scale (CD-RISC)," *Depression and Anxiety* 18 (2003), https://doi.org/10.1002/da.10113.

14. Julian Agyeman, Robert D. Bullard, and Bob Evans, eds., *Just Sustainabilities: Development in an Unequal World* (Cambridge, MA: MIT Press, 2003), 5.

1. THE PROBLEM OF FLOURISHING IN OUR TIMES

1. David Wallace-Wells, *Uninhabitable Earth: Life after Warming* (New York: Tim Duggan Books, 2019).

2. James Howard Kunstler, *The Long Emergency: Surviving the End of Oil, Climate Change, and Other Converging Catastrophes of the Twenty-First Century* (New York: Grove, 2006).

3. Byron Williston, *The Anthropocene Project: Virtue in the Age of Climate Change* (Oxford: Oxford University Press, 2015), 58.

4. Greta Thunberg, *No One Is Too Small to Make a Difference* (London: Penguin, 2019).

5. Michael Shellenberger, *Apocalypse Never: Why Environmental Alarmism Hurts Us All* (New York: Harper, 2020).

6. I use behavior change in a very broad sense, which includes participating in systems change (governance participation, activism, and volunteering), not just beneficial environmental practices like recycling and buying green products. We will need to change systems structures—economic structures, policies, and social norms—in order to enable large-scale behavior changes. I am not suggesting that the combination of lots of individual initiatives is sufficient for addressing our challenges adequately.

7. Robert Ruiter et al., "Sixty Years of Fear Appeal Research: Current State of the Evidence," *International Journal of Psychology* 49 (2014), https://doi.org/10.1002/ijop.12042.

8. Will Steffen et al., "Planetary Boundaries: Guiding Human Development on a Changing Planet," *Science* 347 (2015), https://doi.org/10.1126/science.1259855. See also Johan Rockström et al., "A Safe Operating Space for Humanity," *Nature* 461 (2009), and Will Steffen et al., "Trajectories of the Earth System in the Anthropocene," *Proceedings of the National Academy of Sciences* 115 (2018).

9. Rather than provide references for every challenge described in this section, I summarize key resources for our challenges and their implications in this bibliographic endnote. For a thorough though somewhat dated account of ecological challenges see the Millennium Ecosystem Assessment reports available at https://www.millenniumassessment.org/en/index.html. For more recent assessment of climate change challenges see the IPCC 2023 AR 6 synthesis report, https://www.ipcc.ch/report/ar6/syr/, and the "Fifth National Climate Assessment," https://nca2023.globalchange.gov/. In addition, reports

and books that describe a wide range of challenges we will face in the near future and which have influenced my analysis include the US National Intelligence Council's *Global Trends 2040* (2021), https://www.dni.gov/index.php/gt2040-home; Richard Heinberg, *Welcome to the Great Unraveling: Navigating the Polycrisis of Environmental and Social Breakdown* (Corvallis, OR: Post Carbon Institute, 2023); Bill McKibben, *Falter: Has the Human Game Begun to Play Itself Out?* (New York: Henry Holt, 2019); Naomi Oreskes and Erik Conway, *The Collapse of Western Civilization: A View from the Future* (New York: Columbia University Press, 2014); Alec Ross, *The Industries of the Future* (New York: Simon & Schuster, 2017); and Darrell M. West, *The Future of Work: Robots, AI, and Automation* (Washington, DC: Brookings Institution, 2018).

10. Glenn Albrecht et al., "Solastalgia: The Distress Caused by Environmental Change," *Australasian Psychiatry: Bulletin of Royal Australian and New Zealand College of Psychiatrists* 15 (2007), https://doi.org/10.1080/10398560701701288.

11. Jennifer Ortman, Victoria Velkoff, and Howard Hogan, "An Aging Nation: The Older Population in the United States," https://www.census.gov/library/publications/2014/demo/p25-1140.html.

12. Horst Rittel and Melvin Webber, "Dilemmas in a General Theory of Planning," *Policy Sciences* 4 (1973), https://doi.org/10.1007/BF01405730.

13. Lance H. Gunderson and C. S. Holling, eds., *Panarchy: Understanding Transformations in Human and Natural Systems* (Washington, DC: Island Press, 2002).

14. Brian D. Fath, Carly A. Dean, and Harald Katzmair, "Navigating the Adaptive Cycle: An Approach to Managing the Resilience of Social Systems," *Ecology and Society* 20 (2015), https://doi.org/10.5751/ES-07467-200224.

15. The study of civilization collapse might be used to bolster apocalyptic interpretations. See for example Jared Diamond, *Collapse: How Societies Choose to Fail or Succeed*, rev. ed. (New York: Penguin Books, 2011), and Robert Costanza, Lisa Graumlich, and W. Steffen, *Sustainability or Collapse? An Integrated History and Future of People on Earth* (Cambridge, MA: MIT Press, 2007). We need more research on cultural transformation after release to balance our picture of potential outcomes.

16. See Richard Heinberg, "The Big Picture," https://www.resilience.org/stories/2018-12-17/the-big-picture/, for an eloquent argument that we are in the late conservation stage of the adaptive cycle.

17. Timothy M. Lenton et al., "Tipping Elements in the Earth's Climate System," *Proceedings of the National Academy of Sciences of the United States of America* 105 (2008), https://doi.org/10.1073/pnas.0705414105.

18. A notable exception to my generalization about failures of international organization to keep us away from key thresholds was the creation of the Montreal Protocol that created a very effective international agreement to reduce use of chlorofluorocarbons and thereby avoid further decline of the ozone layer.

19. "Looking to the Future, Public Sees an America in Decline on Many Fronts," Travis Mitchell, *Pew Research Center's Social & Demographic Trends Project* (blog), March 21, 2019, https://www.pewresearch.org/social-trends/2019/03/21/public-sees-an-america-in-decline-on-many-fronts/.

20. Fath, Dean, and Katzmair, "Navigating the Adaptive Cycle."

21. Brian Walker and David Salt, *Resilience Practice: Building Capacity to Absorb Disturbance and Maintain Function* (Washington, DC: Island Press, 2012), 3.

22. Fath, Dean, and Katzmair, "Navigating the Adaptive Cycle," 24.

23. Annie Gowen, "The Town That Built Back Green," *Washington Post*, October 23, 2020, https://www.washingtonpost.com/climate-solutions/2020/10/22/greensburg-kansas-wind-power-carbon-emissions/.

24. Kim Stanley Robinson, *The Ministry for the Future* (New York: Orbit, 2020).

2. COLLABORATING WELL IN A COMPETITIVE CULTURE

1. Barack Obama, speech to Planned Parenthood Action Fund, filmed on July 17, 2007 (video of part of speech, at 1:43), https://www.youtube.com/watch?v=lXFrAvoO3vk.

2. Kathryn Abrams, "Empathy and Experience in the Sotomayor Hearings," *Ohio Northern University Law Review* 36 (2010).

3. July 13, 2009, Judiciary Committee hearings, quoted in Abrams, "Empathy and Experience," 267.

4. Sonia Sotomayor, "A Latina Judge's Voice," *Berkeley La Raza Law Journal* 13 (2002): 92.

5. Paul Bloom, *Against Empathy: The Case for Rational Compassion* (New York: Ecco, 2016).

6. Joshua Rosenberg, "Teaching Empathy in Law School," *University of San Francisco Law Review* 36 (2002), https://repository.usfca.edu/usflawreview/vol36/iss3/4; Martin Hoffman, "Empathy, Justice, and the Law," in *Empathy: Philosophical and Psychological Perspectives*, ed. Amy Coplan and Peter Goldie (New York: Oxford University Press, 2011), 230.

7. Barbara Gray, "Conditions Facilitating Interorganizational Collaboration," *Human Relations* 38 (1985): 911–36, https://doi.org/10.1177/001872678503801001.

8. Bryan G. Norton, *Sustainable Values, Sustainable Change: A Guide to Environmental Decision Making* (Chicago: University of Chicago Press, 2015).

9. Deborah Tannen, *The Argument Culture: Stopping America's War of Words* (New York: Ballantine Books, 1999).

10. Daniel Markovits, *The Meritocracy Trap: How America's Foundational Myth Feeds Inequality, Dismantles the Middle Class, and Devours the Elite* (New York: Penguin, 2019).

11. Fath, Dean, and Katzmair, "Navigating the Adaptive Cycle."

12. "Trust and Distrust in America," Pew Research Center, https://www.people-press.org/2019/07/22/the-state-of-personal-trust/.

13. Michael Shellenberger and Ted Nordhouse, "The Death of Environmentalism: Global Warming Politics in a Post Environmental World" (2004), www.thebreakthrough.org/images/Death_of_Environmentalism.pdf.

14. Norton, *Sustainable Values, Sustainable Change*.

15. Julia Marie Wondolleck and Steven Lewis Yaffee, *Making Collaboration Work: Lessons from Innovation in Natural Resource Management* (Washington, DC: Island Press, 2000).

16. Paul Sabatier et al., eds., *Swimming Upstream: Collaborative Approaches to Watershed Management* (Cambridge, MA: MIT Press, 2005).

17. Sonia Sotomayor, *My Beloved World*, repr. ed. (New York: Vintage Books, 2014), 126.

18. Heather Battaly, "Is Empathy a Virtue?," in *Empathy: Philosophical and Psychological Perspectives*, ed. Amy Coplan and Peter Goldie (New York: Oxford University Press), 277–301.

19. Abrams, "Empathy and Experience."

20. Bloom, *Against Empathy*; Jesse Prinz, "Against Empathy," *Southern Journal of Philosophy* 49 (2011), https://doi.org/10.1111/j.2041-6962.2011.00069.x.

21. Robert C. Solomon, *The Passions* (Garden City, NY: Anchor / Doubleday, 1976); Martha C. Nussbaum, *Upheavals of Thought: The Intelligence of Emotions* (Cambridge: Cambridge University Press, 2001); Antonio Damasio, *Descartes' Error: Emotion, Reason, and the Human Brain*, repr. ed. (London: Penguin Books, 2005).

22. Hilary Putnam, *The Many Faces of Realism* (La Salle, IL: Open Court, 1987).

23. Sotomayor, "Latina Judge's Voice."

24. Abrams, "Empathy and Experience," 271.

25. This section of the chapter is adapted from Paul Stonehouse and William Throop, "Coping with Climate Despair: Cultivating the Skills of Hope and Tranquil Resolve," *Journal of Sustainability Education* 28 (2023).

26. Adrienne Martin, *How We Hope: A Moral Psychology* (Princeton, NJ: Princeton University Press, 2013).

27. Victoria McGeer, "The Art of Good Hope," *Annals of the American Academy of Political and Social Science* 592 (2004).

28. Paul Hawken, *Blessed Unrest: How the Largest Social Movement in History Is Restoring Grace, Justice, and Beauty to the World*, repr. ed. (New York: Penguin Books, 2008), 4.

29. Williston, *Anthropocene Project*.

30. Hawken, *Blessed Unrest*, 4.

31. C. R. Snyder, "Conceptualizing, Measuring, and Nurturing Hope," *Journal of Counseling & Development* 73 (1995): 357–58, https://doi.org/10.1002/j.1556-6676.1995.tb01764.x.

32. McGeer, "Art of Good Hope."

33. Martin Luther King Jr., "I Have a Dream," speech presented at the March on Washington for Jobs and Freedom, Washington, DC, August 26, 1963, http://avalon.law.yale.edu/20th_century/mlk01.asp.

34. Jonathan Lear, *Radical Hope: Ethics in the Face of Cultural Devastation* (Cambridge, MA: Harvard University Press, 2008).

35. Williston, *Anthropocene Project*; Allen Thompson, "Radical Hope for Living Well in a Warmer World," *Journal of Agricultural and Environmental Ethics* 23 (2009), https://doi.org/10.1007/s10806-009-9185-2.

36. Robert Putnam, *Bowling Alone: The Collapse and Revival of American Community* (New York: Simon & Schuster, 2000).

37. Robert C. Solomon and Fernando F. Flores, *Building Trust: In Business, Politics, Relationships, and Life* (New York: Oxford University Press, 2001).

38. Solomon and Flores, 64.

39. McGeer, "Art of Good Hope."

40. Robert D. Putnam and Shaylyn Romney Garrett, *The Upswing: How America Came Together a Century Ago and How We Can Do It Again* (New York: Simon & Schuster, 2020). Putnam and Garrett use the General Social Survey, as well as other surveys and demographic datasets, to provide measures of different components of social capital. They triangulate among multiple questions to make claims about the decline of trust.

41. Gallup, "Confidence in Institutions," https://news.gallup.com/poll/1597/confidence-institutions.aspx.

42. Solomon and Flores, *Building Trust*, 57.

43. Roy J. Lewicki and Chad Brinsfield, "Trust Repair," *Annual Review of Organizational Psychology and Organizational Behavior* 4 (2017).

44. Roderick M. Kramer and Roy J. Lewicki, "Repairing and Enhancing Trust: Approaches to Reducing Organizational Trust Deficits," *Academy of Management Annals* 4 (2010).

45. Sotomayor, *My Beloved World*, 91.

46. Sotomayor, 138.

47. Sotomayor, 226.

48. American Association of Schools and Colleges, "Fulfilling the American Dream: Liberal Education and the Future of Work," https://www.aacu.org/sites/default/files/files/LEAP/2018EmployerResearchReport.pdf.

49. Sotomayor, *My Beloved World*, 356.

50. Sotomayor, 357.

3. RECOVERING HUMILITY AND SOFTENING CONVICTION

1. Robert T. Pennock, *An Instinct for Truth: Curiosity and the Moral Character of Science* (Cambridge, MA: MIT Press, 2019).

2. David Barker, Ryan Detamble, and Morgan Marietta, "Intellectualism, Anti-intellectualism, and Epistemic Hubris in Red and Blue America," *American Political Science Review* 116 (2021), https://doi.org/10.1017/S0003055421000988.

3. Michael P. Lynch, *Know-It-All Society: Truth and Arrogance in Political Culture* (New York: Liveright, 2019).

4. Lynch, 57.

5. Robert P. Abelson, "Conviction," *American Psychologist* 43, no. 4 (1988), https://doi.org/10.1037/0003-066X.43.4.267.

6. Michael D. Slater, "Reinforcing Spirals: The Mutual Influence of Media Selectivity and Media Effects and Their Impact on Individual Behavior and Social Identity," *Communication Theory* 17, no. 3 (2007): 281–303, https://doi.org/10.1111/j.1468-2885.2007.00296.x.

7. Cass R. Sunstein, "The Law of Group Polarization," *Journal of Political Philosophy* 10 (2002): 175–95, https://doi.org/10.1111/1467-9760.00148.

8. Raymond S. Nickerson, "Confirmation Bias: A Ubiquitous Phenomenon in Many Guises," *Review of General Psychology* 2 (1998), https://doi.org/10.1037/1089-2680.2.2.175.

9. P. C. Wason, "On the Failure to Eliminate Hypotheses in a Conceptual Task," *Quarterly Journal of Experimental Psychology* 12 (1960), https://doi.org/10.1080/17470216008416717.

10. William Butler Yeats, "The Second Coming," Poetry Foundation, 1919, https://www.poetryfoundation.org/poems/43290/the-second-coming.

11. Lynch, *Know-It-All Society*, 23.

12. Steve Schwartz pointed out to me that this kind of account might imply that many fundamentalist religious people are intellectually arrogant. Having strong religious beliefs that are part of one's identity is not sufficient for arrogance, but if one adds having epistemic certainty about those beliefs and the view that all other religions are mistaken, then one approaches arrogance. Sometimes people confuse strong emotional commitment with certainty, but in my experience once we draw this distinction clearly, many fundamentalists espouse commitment rather than certainty regarding their religious beliefs. To be clear, having some convictions can be good.

13. In drawing the hard/soft conviction distinction, I am influenced by Imre Lakatos's distinction between the hard core of a scientific research program, which is unmodifiable, and the soft core, which may be altered. I am not suggesting, however, that Lakatos's distinction is a matter of degree of conviction. Imre Lakatos, "Falsification and the Methodology of Scientific Research Programmes," in *Criticism and the Growth of Knowledge*, ed. Imre Lakatos and Alan Musgrave (Cambridge: Cambridge University Press, 1970), 91–196, https://doi.org/10.1017/CBO9781139171434.009.

14. Kenny Walker and Lynda Walsh, "'No One Yet Knows What the Ultimate Consequences May Be': How Rachel Carson Transformed Scientific Uncertainty into a Site for Public Participation in *Silent Spring*," *Journal of Business and Technical Communication* 26, no. 1 (January 1, 2012), https://doi.org/10.1177/1050651911421122.

15. Rachel Carson, *Silent Spring*, 40th anniversary ed. (Boston: Mariner Books, 2002), quotations respectively on 23, 211, 211, 22.

16. Walker and Walsh, "No One Yet Knows," 25–26.

17. Walker and Walsh, 26–27.

18. Dennis Whitcomb et al., "Intellectual Humility: Owning Our Limitations," *Philosophy and Phenomenological Research* 94, no. 3 (2017), https://doi.org/10.1111/phpr.12228.

19. Whitcomb et al., "Intellectual Humility."

20. Todd Kashdan et al., "The Five-Dimensional Curiosity Scale: Capturing the Bandwidth of Curiosity and Identifying Four Unique Subgroups of Curious People," *Journal of Research in Personality* 73 (2018), https://doi.org/10.1016/J.JRP.2017.11.011.

21. See for example Jim Collins, "Level 5 Leadership: The Triumph of Humility and Fierce Resolve," *Harvard Business Review* 79 (2001).

22. William Throop and Matt Mayberry, "Leadership for the Sustainability Transition," *Business and Society Review* 122 (2017), https://doi.org/10.1111/basr.12116.

23. Robert K. Greenleaf, *Servant Leadership: A Journey into the Nature of Legitimate Power and Greatness* (New York: Paulist, 1977).

24. Robert Emmons and Michael McCullough, eds., *The Psychology of Gratitude*, Series in Affective Science (Oxford: Oxford University Press, 2004).

25. Everett Worthington, Don E. Davis, and Joshua N. Hook, *Handbook of Humility: Theory, Research, and Applications* (New York: Routledge, 2016); Robert C. Roberts and W. Jay Wood, *Intellectual Virtues: An Essay in Regulative Epistemology* (Oxford: Oxford University Press, 2007), https://doi.org/10.1093/9780199283675.001.0001.

26. Robert Jonathan Cabin, *Intelligent Tinkering: Bridging the Gap between Science and Practice* (Washington, DC: Island Press, 2011).

27. Aldo Leopold, *A Sand County Almanac*, 1st Ballantine ed. (New York: Sierra Club / Ballantine, 1970), 190.

28. Cabin, *Intelligent Tinkering*, 199.

29. Aristotle, *Aristotle's Nicomachean Ethics*.

30. Aristotle's views on the role of friendship in character building are quite complex. For an excellent treatment of this topic see Diana Hoyos-Valdés, "The Notion of Character Friendship and the Cultivation of Virtue," *Journal for the Theory of Social Behavior* 48 (2017): 66–82.

31. Rachel Carson and Dorothy Freeman, *Always, Rachel: The Letters of Rachel Carson and Dorothy Freeman, 1952–1964*, ed. Martha E. Freeman (Boston: Beacon, 1995), 20.

32. Robert Cabin, personal communication with the author, January 24, 2020.

4. TECHNOLOGICAL FANTASY AND FRUGALITY

1. Parts of this chapter are adapted from William Throop, "Frugality and Resilience: A Pragmatist Meditation," in *Pragmatist and American Philosophical Perspectives on Resilience*, ed. Kelly Parker and Heather Keith (New York: Lexington Books, 2019), and used with permission.

2. Donella Meadows et al., *The Limits to Growth: A Report for the Club of Rome's Project on the Predicament of Mankind* (New York: Universe Books, 1972).

3. Techno-optimists include Andrew McAfee, Steven Pinker, Christine Lagarde, Eric Schmidt, and Larry Summers.

4. Lydia Saad, "Socialism as Popular as Capitalism among Young Adults in U.S.," Gallup, https://news.gallup.com/poll/268766/socialism-popular-capitalism-among-young-adults.aspx.

5. Joshua Yates and James Davison Hunter, eds., *Thrift and Thriving in America: Capitalism and Moral Order from the Puritans to the Present* (New York: Oxford University Press, 2011), 3.

6. David Bentley Hart, *The Hidden and the Manifest: Essays in Theology and Metaphysics* (Grand Rapids, MI: Eerdmans, 2017).

7. Peterson and Seligman, *Character Strengths*. They do have a chapter on self-regulation, but the chapter does not address consumptive practices, which give rise to the importance of frugality as a resilience virtue.

8. Terrence H. Witkowski, "A Brief History of Frugality Discourses in the United States," *Consumption Markets & Culture* 13 (2010), https://doi.org/10.1080/10253861003786975.

9. Bruce Piasecki, in *Doing More with Less: The New Way to Wealth* (Hoboken, NJ: Wiley, 2012), makes a strong case for Franklin's being the exemplar that business needs to motivate frugality.

10. Benjamin Franklin, *Autobiography and Other Writings*, ed. Ormand Seavey (Oxford: Oxford University Press, 2009), 81.

11. Most of these proverbs were collected in Franklin's *The Way to Wealth*, originally published in the 1758 issue of *Poor Richard's Almanac*. Some were borrowed from earlier authors and later used in more familiar forms, like "A penny saved is a penny earned."

12. See Walter Isaacson, *Benjamin Franklin: An American Life* (New York: Simon & Schuster, 2004). Franklin's motivations for his own early frugality were undoubtedly mixed. Though his published advice often focused on the self-restraint necessary to achieve later wealth, he was also deeply influenced by his Quaker friends in Philadelphia whose motivations for frugal lifestyles were more spiritual and in line with constructive frugality.

13. W. Mischel, E. B. Ebbesen, and A. R. Zeiss, "Cognitive and Attentional Mechanisms in Delay of Gratification," *Journal of Personality and Social Psychology* 21 (1972), https://doi.org/10.1037/h0032198.

14. Walter Mischel, Yuichi Shoda, and Philip K. Peake, "The Nature of Adolescent Competencies Predicted by Preschool Delay of Gratification," *Journal of Personality and Social Psychology* 54 (1988), https://doi.org/10.1037/0022-3514.54.4.687.

15. Peterson and Seligman, *Character Strengths*.

16. Rachel Kaplan and Ruby K. Blume, *Urban Homesteading: Heirloom Skills for Sustainable Living* (New York: Skyhorse, 2011).

17. Robin Wall Kimmerer, *Braiding Sweetgrass: Indigenous Wisdom, Scientific Knowledge, and the Teachings of Plants* (Minneapolis: Milkweed, 2020).

18. The increasing use of school gardens to teach science and nutrition is a notable exception; see chapter 7 for more detail on this important development.

19. Luk Bouckaert, Hendrik Opdebeeck, and László Zsolnai, *Frugality: Rebalancing Material and Spiritual Values in Economic Life* (New York: Peter Lang, 2008).

20. Juliet Schor, *Plenitude: The New Economics of True Wealth* (New York: Penguin, 2010).

21. Epicurus, "Letter to Menoeceus," in *The Stoic and Epicurean Philosophers: The Complete Extant Writings of Epicurus, Epictetus, Lucretius [and] Marcus Aurelius*, ed. Whitney Jennings Oates (New York: Random House, 1940).

22. Edward Abbey, *Beyond the Wall: Essays from the Outside* (New York: Holt, Rinehart and Winston, 1984), 153. For more on finding beauty in nature see Annie Dillard, *Pilgrim at Tinker Creek* (New York: Harper Perennial Modern Classics, 2013), and Terry Tempest Williams, *Finding Beauty in a Broken World* (New York: Vintage Books, 2009). Allen Carlson's *Aesthetics and the Environment: The Appreciation of Nature, Art and Architecture* (New York: Routledge, 1999) is an excellent philosophical treatment of the topic.

23. Williams, *Finding Beauty*.

24. In an earlier paper on frugality, I identified a third kind of frugality—integral frugality—which I identified with touching the world lightly: William Throop, "Frugality and Resilience: A Pragmatist Meditation," in *Pragmatist and American Philosophical Perspectives on Resilience*, ed. Kelly Parker and Heather Keith (New York: Lexington Books, 2019). In the current chapter it seems clearer to combine constructive and integral frugality.

25. Lao Tzu, *Tao Te Ching*, trans. R. B. Blakney (New York: Signet, 2007).

26. Emrys Westacott, *The Wisdom of Frugality: Why Less Is More—More or Less* (Princeton, NJ: Princeton University Press, 2016).

27. Fath, Dean, and Katzmair, "Navigating the Adaptive Cycle."

28. Walker and Salt, *Resilience Practice*.

29. "Real Personal Consumption Expenditures [PCECCA]," US Bureau of Economic Analysis, retrieved from FRED, Federal Reserve Bank of Saint Louis, https://fred.stloui sfed.org/series/PCECCA.

30. Mark J. Perry, "New US Homes Today Are 1,000 Square Feet Larger Than in 1973 and Living Space per Person Has Nearly Doubled," American Enterprise Institute, https://www.aei.org/carpe-diem/new-us-homes-today-are-1000-square-feet-larger-than-in-1973-and-living-space-per-person-has-nearly-doubled/.

31. Max Roser, Hannah Ritchie, and Edouard Mathieu, "Technological Change" (2013), published online at OurWorldInData.org, https://ourworldindata.org/techno logical-progress.

32. Jean Twenge et al., "Age, Period, and Cohort Trends in Mood Disorder Indicators and Suicide-Related Outcomes in a Nationally Representative Dataset, 2005–2017," *Journal of Abnormal Psychology* 128 (2019), https://doi.org/10.1037/abn0000410; NORC, "Historic Shift in Americans' Happiness amid Pandemic," 2020, https://www.norc.org/PDFs/COVID%20Response%20Tracking%20Study/Historic%20Shift%20in%20Ameri cans%20Happiness%20Amid%20Pandemic.pdf.

33. Helga Dittmar et al., "The Relationship between Materialism and Personal Well-Being: A Meta-analysis," *Journal of Personality and Social Psychology* 107 (2014), https://doi.org/10.1037/a0037409.

34. Shimon Saphire-Bernstein and Shelley E. Taylor, "Close Relationships and Happiness," in *The Oxford Handbook of Happiness*, ed. Susan David, Ilona Boniwell, and Amanda Conley Ayers (Oxford: Oxford University Press, 2014).

35. Robert Waldinger, "What Makes a Good Life: Lessons from the Longest Study on Happiness," TED talk, 2015, https://www.adultdevelopmentstudy.org/.

36. Schor, *Plenitude*.

37. Bill McKibben, *Maybe One: A Case for Smaller Families* (New York: Plume, 1999).

38. John R. Ehrenfeld, "The Roots of Sustainability," *MIT Sloan Management Review*, January 15, 2005, 23.

39. Schor, *Plenitude*.

40. David A. Crocker and Toby Linden, eds., *Ethics of Consumption: The Good Life, Justice, and Global Stewardship* (Lanham, MD: Rowman & Littlefield, 1997); Peter Singer, *One World: The Ethics of Globalization* (New Haven, CT: Yale University Press, 2002); Clive Barnett, Philip Cafaro, and Terry Newholm, "Philosophy and Ethical Consumption," in *The Ethical Consumer*, ed. Rob Harrison, Terry Newholm, and Deirdre Shaw (London: Sage, 2005).

41. "U.S. Environmental Footprint Factsheet," Center for Sustainable Systems, University of Michigan, pub. no. CSS08–08, http://css.umich.edu/sites/default/files/U.S.%20 Environmental%20Footprint_CSS08-08_e2020.pdf.

42. Andrew McAfee, *More from Less: The Surprising Story of How We Learned to Prosper Using Fewer Resources—and What Happens Next* (New York: Scribner, 2019).

43. Willard Cochrane, *Farm Prices: Myth and Reality* (Minneapolis: University of Minnesota Press, 1958).

44. Andrea Nightingale et al., "Beyond Technical Fixes: Climate Solutions and the Great Derangement," *Climate and Development* 12 (2020), https://doi.org/10.1080/1756 5529.2019.1624495. This article explores in more detail how our focus on technological fixes prevents us from seeing deeper changes that we must make, such as reframing our relation to nature and highlighting the social justice issues that make our current responses untenable.

45. Steven Fesmire, *John Dewey and Moral Imagination: Pragmatism and Ethics* (Bloomington: Indiana University Press, 2003).

46. Franklin, *Autobiography*.

47. Sotomayor, *My Beloved World*.

5. LEARNING TO THINK LIKE A MOUNTAIN

1. Terence J. Centner, *America's Blame Culture: Pointing Fingers and Shunning Restitution* (Durham, NC: Carolina Academic Press, 2008).

2. Nathanael J. Fast and Larissa Z. Tiedens, "Blame Contagion: The Automatic Transmission of Self-Serving Attributions," *Journal of Experimental Social Psychology* 46 (2010), https://doi.org/10.1016/j.jesp.2009.10.007.

3. Leopold, *Sand County Almanac*.

4. John Holland, "Studying Complex Adaptive Systems," *Journal of Systems Science and Complexity* 19 (2006), https://doi.org/10.1007/s11424-006-0001-z.

5. I recommend to the interested reader four excellent books that introduce systems thinking to a nontechnical audience: Peter M. Senge, a prominent management consultant, wrote the classic application of systems theory to business, *The Fifth Discipline: The Art and Practice of the Learning Organization*, rev. and updated ed. (New York: Doubleday, 2006). Donella Meadows, known for her early work on the limits to growth, published a distillation of her understanding of systems thinking in *Thinking in Systems: A Primer*, ed. Diana Wright (White River Junction, VT: Chelsea Green, 2008). Brian Walker and David Salt apply systems thinking to the resilience of socio-ecological systems in *Resilience Practice* (Washington, DC: Island Press, 2012). David Stroh provides an admirably clear guidebook for applying systems thinking to social problems in *Systems Thinking for Social Change: A Practical Guide to Solving Complex Problems, Avoiding Unintended Consequences, and Achieving Lasting Results* (White River Junction, VT: Chelsea Green, 2015). Also, for a very strong review article on systems-thinking skills see Ross D. Arnold and Jon P. Wade, "A Complete Set of Systems Thinking Skills," *Insight* 20 (2017), https://doi.org/10.1002/inst.12159.

6. Meadows, *Thinking in Systems*, 97.

7. Walker and Salt, *Resilience Practice*, 37.

8. Stroh, *Systems Thinking for Social Change*, chap. 4.

9. Kim A. Kastens et al., "How Geoscientists Think and Learn," *Eos, Transactions, American Geophysical Union* 90 (2009), https://doi.org/10.1029/2009EO310001.

10. Jim Collins, *Good to Great: Why Some Companies Make the Leap and Others Don't* (New York: HarperBusiness, 2001).

11. For an excellent discussion of kinds of thresholds and how we can improve our ability to identify them see Walker and Salt, *Resilience Practice*, chap. 1. See also Brian Walker and Jacqueline Meyers, "Thresholds in Ecological and Social-Ecological Systems: A Developing Database," *Ecology and Society* 9 (2004).

12. Niklas Boers, "Observation-Based Early-Warning Signals for a Collapse of the Atlantic Meridional Overturning Circulation," *Nature Climate Change* 11 (2021), https://doi.org/10.1038/s41558-021-01097-4.

13. Meadows, *Thinking in Systems*, chap. 6.

14. Meadows, 146. For a discussion of weak and strong leverage points for the sustainability transition see David J. Abson et al., "Leverage Points for Sustainability Transformation," *Ambio* 46 (2017), https://doi.org/10.1007/s13280-016-0800-y.

15. Leopold, *Sand County Almanac*, 262.

16. In the history of the social sciences, the classical debate between Émile Durkheim and Max Weber reflects the competing approaches to understanding groups. In ecology, there has been a much greater focus on holist explanations, though early debates between Frederic Clements and Henry Gleason regarding plant associations mirror some aspects of the social science debates.

17. Putnam and Garrett, *Upswing*.

18. Putnam and Garrett acknowledge numerous complications in their simplified picture of the I-We-I arc. For example, Black people and women were largely excluded from the early twentieth-century "We." The book's broad generalizations about social norms

apply only to the aggregate population, not to all subpopulations. Michael McGerr further complicates the picture of individualism during the Progressive Era. He notes that the wealthy were highly influenced by individualism, while the urban poor never had the privilege that enabled individualism. Poor people needed a community focus to survive. Michael McGerr, *A Fierce Discontent: The Rise and Fall of the Progressive Movement in America, 1870–1920* (Oxford: Oxford University Press, 2005). Any claim about the dominance of individualism today also requires careful qualification, since emphasis on elements of individualism varies considerably across different social groups.

19. Andrew Dobson, "Environment Sustainabilities: An Analysis and a Typology," *Environmental Politics* 5 (1996): 401–28, https://doi.org/10.1080/09644019608414280.

20. Heather M. Farley and Zachary A. Smith, *Sustainability: If It's Everything, Is It Nothing?*, 2nd ed. (New York: Routledge, 2020).

21. Simon Kirchin, ed., *Thick Concepts*, Mind Association Occasional Series (Oxford: Oxford University Press, 2013).

22. Julian Agyeman, "Toward a 'Just' Sustainability?," *Continuum* 22 (2008), https://doi.org/10.1080/10304310802452487; Julian Agyeman, *Introducing Just Sustainabilities: Policy, Planning, and Practice* (London: Zed Books, 2013).

23. Scott D. Campbell, "Sustainable Development and Social Justice: Conflicting Urgencies and the Search for Common Ground in Urban and Regional Planning," *Michigan Journal of Sustainability* 1 (2013), https://doi.org/10.3998/mjs.12333712.0001.007.

24. Sander van der Leeuw and Carl Folke, "The Social Dynamics of Basins of Attraction," *Ecology and Society* 26 (2021), https://doi.org/10.5751/ES-12289-260133.

25. Rob Hopkins, *The Transition Handbook: From Oil Dependency to Local Resilience* (Cambridge: Green Books, 2014).

6. BARRIERS AND STEPLADDERS

1. Noah J. Webster, Kristine J. Ajrouch, and Toni C. Antonucci, "Sociodemographic Differences in Humility: The Role of Social Relations," *Research in Human Development* 15 (2018), https://doi.org/10.1080/15427609.2017.1414670.

2. Angelos Stamos, Efthymios Altsitsiadis, and Siegfried Dewitte, "Investigating the Effect of Childhood Socioeconomic Background on Interpersonal Trust: Lower Childhood Socioeconomic Status Predicts Lower Levels of Trust," *Personality and Individual Differences* 145 (2019), https://doi.org/10.1016/j.paid.2019.03.011.

3. Michael W. Kraus, Stéphane Côté, and Dacher Keltner, "Social Class, Contextualism, and Empathic Accuracy," *Psychological Science* 21 (2010), https://doi.org/10.1177/0956797610387613.

4. Igor Grossmann and Justin Brienza, "The Strengths of Wisdom Provide Unique Contributions to Improved Leadership, Sustainability, Inequality, Gross National Happiness, and Civic Discourse in the Face of Contemporary World Problems," *Journal of Intelligence* 6 (2018): 22, https://doi.org/10.3390/jintelligence6020022.

5. Justin P. Brienza and Igor Grossmann, "Social Class and Wise Reasoning about Interpersonal Conflicts across Regions, Persons and Situations," *Proceedings of the Royal Society B: Biological Sciences* 284 (2017), https://doi.org/10.1098/rspb.2017.1870.

6. I am indebted to Steve Schwartz for advancing a powerful version of this concern (personal communication, July 26, 2021).

7. Martha C. Nussbaum, *The Fragility of Goodness: Luck and Ethics in Greek Tragedy and Philosophy*, 2nd ed. (Cambridge: Cambridge University Press, 2001).

8. "Senator Hawley Delivers National Conservatism Keynote on the Left's Attack on Men in America," Senator Josh Hawley, https://www.hawley.senate.gov/senator-hawley-delivers-national-conservatism-keynote-lefts-attack-men-america.

9. Janet S. Hyde, Elizabeth Fennema, and Susan J. Lamon, "Gender Differences in Mathematics Performance: A Meta-analysis," *Psychological Bulletin* 107 (1990), https://doi.org/10.1037/0033-2909.107.2.139.

10. Ashley E. Thompson and Daniel Voyer, "Sex Differences in the Ability to Recognize Non-verbal Displays of Emotion: A Meta-analysis," *Cognition and Emotion* 28 (2014), https://doi.org/10.1080/02699931.2013.875889.

11. Daniel Lerch, *Six Foundations for Building Community Resilience* (Corvallis, OR: Post Carbon Institute, 2015).

12. Angela Duckworth, *Grit: The Power of Passion and Perseverance* (New York: Scribner, 2018).

13. Reeves and Deimler argue that adaptability is the most promising new source of advantage in business competition. Martin Reeves and Mike Deimler, "Adaptability: The New Competitive Advantage," *Harvard Business Review*, https://hbr.org/2011/07/adaptability-the-new-competitive-advantage.

14. Michel de Montaigne, "Of Presumption," in *The Essays* (1580), quoted in and translated by Ward Farnsworth in his *The Practicing Stoic: A Philosophical User's Manual* (Jaffrey, NH: David R. Godine, 2018), 26.

15. Hawken, *Blessed Unrest*.

16. Andreas Karelas, *Climate Courage: How Tackling Climate Change Can Build Community, Transform the Economy, and Bridge the Political Divide in America* (Boston: Beacon, 2020).

17. Putnam and Garrett, *Upswing*, summarized in chapter 5.

18. "Positive Character Traits Education," Texas Education Agency, https://tea.texas.gov/academics/learning-support-and-programs/positive-character-traits-education.

19. This section is adapted from William Throop, "Flourishing in the Age of Climate Change: Finding the Heart of Sustainability," *Midwest Studies in Philosophy* 40 (2016), https://doi.org/10.1111/misp.12062.

20. Valerie Tiberius, "Value Commitments and the Balanced Life," *Utilitas* 17 (2005): 28, https://doi.org/10.1017/s0953820804001384.

7. EDUCATION AND CULTURE CHANGE

1. Parts of this chapter are adapted from William Throop, "Learning Our Way toward Resilience," in *The Community Resilience Reader: Essential Resources for an Era of Upheaval*, ed. Daniel Lerch (Washington, DC: Island Press, 2017) and used with permission.

2. Anthony Biglan and Dennis D. Embry, "A Framework for Intentional Cultural Change," *Journal of Contextual Behavioral Science* 2 (2013), https://doi.org/10.1016/j.jcbs.2013.06.001; David Sloan Wilson et al., "Evolving the Future: Toward a Science of Intentional Change," *Behavioral and Brain Sciences* 37 (2014), https://doi.org/10.1017/S0140525X13001593.

3. For an insightful account of different kinds of educational programs see Donald Mocker and George Spear, "Lifelong Learning: Formal, Nonformal, Informal, and Self-Directed," Information Series 241 (Columbus, OH: National Center, 1982).

4. Tim Kautz et al., "Fostering and Measuring Skills: Improving Cognitive and Non-cognitive Skills to Promote Lifetime Success," OECD Education Working Papers 110 (2014), https://doi.org/10.1787/5jxsr7vr78f7-en.

5. James J. Heckman and Tim Kautz, "Hard Evidence on Soft Skills," *Labour Economics* 19 (2012), https://doi.org/10.1016/j.labeco.2012.05.014.

6. Lorea Martínez and Hanna Melnick, "How One Elementary School Integrates Social-Emotional Skills in the Classroom," *Greater Good Magazine*, May 21, 2019, https://greatergood.berkeley.edu/article/item/how_one_elementary_school_integrates_social_emotional_skills_in_the_classro.

7. "What Is the CASEL Framework?," CASEL, https://casel.org/fundamentals-of-sel/what-is-the-casel-framework/.

8. Joseph A. Durlak et al., "The Impact of Enhancing Students' Social and Emotional Learning: A Meta-analysis of School-Based Universal Interventions," *Child Development* 82 (2011), https://doi.org/10.1111/j.1467-8624.2010.01564.x.

9. "How Learning Happens: Supporting Students' Social, Emotional, and Academic Development," Aspen Institute, 5, https://www.aspeninstitute.org/publications/learning-happens-supporting-students-social-emotional-academic-development/.

10. Claire Lampen, "What Is the Conservative Beef with 'Social-Emotional Learning'?," *Cut*, April 19, 2022, https://www.thecut.com/2022/04/conservative-backlash-social-emotional-learning.html.

11. Paul Tough, "What If the Secret to Success Is Failure?," *New York Times*, September 14, 2011, https://www.nytimes.com/2011/09/18/magazine/what-if-the-secret-to-success-is-failure.html.

12. Marvin Berkowitz and Melinda Bier, "Research-Based Character Education," *Annals of the American Academy of Political and Social Science* 591 (2004), https://doi.org/10.1177/0002716203260082.

13. James J. Heckman and Tim Kautz, "Fostering and Measuring Skills: Interventions That Improve Character and Cognition," NBER Working Papers (2013), https://ideas.repec.org/p/nbr/nberwo/19656.html; Durlak et al., "Impact."

14. "UN Decade of ESD," UNESCO, last modified August 9, 2018, https://en.unesco.org/themes/education-sustainable-development/what-is-esd/un-decade-of-esd.

15. Kate G. Burt et al., "School Gardens in the United States: Current Barriers to Integration and Sustainability," *American Journal of Public Health* 108 (2018), https://doi.org/10.2105/AJPH.2018.304674.

16. For more information about green schools see "The Teaching Building: Current Practices in Sustainability in the 21st Century Classroom," Green Schools National Network, https://greenschoolsnationalnetwork.org/teaching-building-current-practices-sustainability 21st century classroom/.

17. One can acquire knowledge about system functioning without any of the skills that this knowledge might motivate. Since most sustainability curricula focus on practices that enable systems to function better, they tend to foster both the development of habits and the skills that enable us to effectively shift systems. Only with practice and evaluation of results do such skillful habits become refined, however.

18. "Mission & Values," Common Ground School, https://commongroundct.org/about/. Information about the Common Ground School came from its website, https://commongroundct.org.

19. "HS-ETS1–4 Engineering Design," Next Generation Science Standards, https://www.nextgenscience.org/pe/hs-ets1-4-engineering-design.

20. National Science Teaching Association, "K–12 Science Standards Adoption," https://ngss.nsta.org/About.aspx.

21. "Character Education in Universities: A Framework for Flourishing," University of Birmingham, Jubilee Centre for Character & Virtues, https://www.jubileecentre.ac.uk/userfiles/jubileecentre/pdf/character-education/Character_Education_in_Universities_Final_01.pdf.

22. Anthony D. Cortese, "The Critical Role of Higher Education in Creating a Sustainable Future," *Planning for Higher Education* 31 (2003): 3.

23. "The Princeton Review Guide to Green Colleges: 2022 Edition Press Release," https://www.princetonreview.com/press/green-guide/press-release-2022.

24. William Throop, "From Environmental Advocates to Sustainability Entrepreneurs: Rethinking a Sustainability Focused General Education Program," in *Sustainability in*

Higher Education: Stories and Strategies for Transformation, ed. Peggy F. Barlett and Geoffrey W. Chase (Cambridge, MA: MIT Press, 2013).

25. Mitch Thomashow, *The Nine Elements of a Sustainable Campus* (Cambridge, MA: MIT Press, 2016).

26. See also "OECD Skills Outlook 2021: Learning for Life," OECD, Paris, https://doi.org/10.1787/0ae365b4-en.

27. Courses on these topics are regularly offered by Coursera at https://www.coursera.org/. The content is typically up to date and reflects current research, though assessment of performance often requires fees and may not be very rigorous.

28. Stephen Sterling, "Learning for Resilience, or the Resilient Learner? Towards a Necessary Reconciliation in a Paradigm of Sustainable Education," *Environmental Education Research* 16 (2010).

29. Putnam, *Bowling Alone*.

30. Norton, *Sustainable Values*.

31. J. T. Woolley, M. V. McGinnis, and J. Kellner, "The California Watershed Movement: Science and the Politics of Place," *Natural Resources Journal* 42 (2002). See also Michael V. McGinnis, *Science and Sensibility: Negotiating an Ecology of Place* (Oakland: University of California Press, 2016).

32. "Average Annual Hours Actually Worked per Worker," OECD.Stat, https://stats.oecd.org/index.aspx?DataSetCode=ANHRS.

33. Throop and Mayberry, "Leadership."

34. Edgar H. Schein, "How Can Organizations Learn Faster? The Challenge of Entering the Green Room," *MIT Sloan Management Review*, January 15, 1993.

35. Katharine Hayhoe, *Saving Us: A Climate Scientist's Case for Hope and Healing in a Divided World* (New York: Atria / One Signal, 2021).

EPILOGUE

1. Brian C. O'Neill et al., "The Roads Ahead: Narratives for Shared Socioeconomic Pathways Describing World Futures in the 21st Century," *Global Environmental Change* 42 (2017), https://doi.org/10.1016/j.gloenvcha.2015.01.004.

2. O'Neill et al., 172. All elements of my descriptions are paraphrases of the original scenario description.

3. From Mary Oliver's poem "It Was Early," in *Evidence: Poems*, repr. ed. (Boston: Beacon, 2010), 20.

4. Friedrich Nietzsche, *The Gay Science: With a Prelude in Rhymes and an Appendix of Songs*, trans. Walter Kaufmann (New York: Vintage Books, 1974), 47. I am grateful to Steve Schwartz for suggesting I use this metaphor and quotation.

Bibliography

Abbey, Edward. *Beyond the Wall: Essays from the Outside*. New York: Holt, Rinehart and Winston, 1984.

Abelson, Robert P. "Conviction." *American Psychologist* 43, no. 4 (1988): 267–75.

Abrams, Kathryn. "Empathy and Experience in the Sotomayor Hearings." *Ohio Northern University Law Review* 36 (2010): 263–86.

Abson, David J., Joern Fischer, Julia Leventon, Jens Newig, Thomas Schomerus, Ulli Vilsmaier, Henrik von Wehrden, et al. "Leverage Points for Sustainability Transformation." *Ambio* 46, no. 1 (2017): 30–39. https://doi.org/10.1007/s13280-016-0800-y.

Agyeman, Julian. *Introducing Just Sustainabilities: Policy, Planning, and Practice*. London: Zed Books, 2013.

Agyeman, Julian. "Toward a 'Just' Sustainability?" *Continuum* 22, no. 6 (2008): 751–56.

Agyeman, Julian, Robert D. Bullard, and Bob Evans, eds. *Just Sustainabilities: Development in an Unequal World*. Cambridge, MA: MIT Press, 2003.

Albrecht, Glenn, Gina-Maree Sartore, Linda Connor, Nick Higginbotham, Sonia Freeman, Brian Kelly, Helen Stain, Anne Tonna, and Georgia Pollard. "Solastalgia: The Distress Caused by Environmental Change." *Australasian Psychiatry: Bulletin of Royal Australian and New Zealand College of Psychiatrists* 15, Suppl. 1 (2007): S95–98.

Aristotle. *Aristotle's Nicomachean Ethics*. Translated by Robert C. Bartlett and Susan D. Collins. Chicago: University of Chicago Press, 2012.

Arnold, Ross D., and Jon P. Wade. "A Complete Set of Systems Thinking Skills." *Insight* 20, no. 3 (2017): 9–17. https://doi.org/10.1002/inst.12159.

Barker, David, Ryan Detamble, and Morgan Marietta. "Intellectualism, Anti-intellectualism, and Epistemic Hubris in Red and Blue America." *American Political Science Review* 116, no. 1 (2021): 38–53. https://doi.org/10.1017/S0003055421000988.

Barnett, Clive, Philip Cafaro, and Terry Newholm. "Philosophy and Ethical Consumption." In *The Ethical Consumer*, edited by Rob Harrison, Terry Newholm, and Deirdre Shaw, 11–24. London: Sage, 2005.

Battaly, Heather. "Is Empathy a Virtue?" In *Empathy: Philosophical and Psychological Perspectives*, edited by Amy Coplan and Peter Goldie, 277–301. New York: Oxford University Press, 2011.

Berkowitz, Marvin, and Melinda Bier. "Research-Based Character Education." *Annals of the American Academy of Political and Social Science* 591 (2004): 72–85.

Biglan, Anthony, and Dennis D. Embry. "A Framework for Intentional Cultural Change." *Journal of Contextual Behavioral Science* 2, no. 3–4 (2013). https://doi.org/10.1016/j.jcbs.2013.06.001.

Bloom, Paul. *Against Empathy: The Case for Rational Compassion*. New York: Ecco, 2016.

Boers, Niklas. "Observation-Based Early-Warning Signals for a Collapse of the Atlantic Meridional Overturning Circulation." *Nature Climate Change* 11, no. 8 (2021): 680–88. https://doi.org/10.1038/s41558-021-01097-4.

Bouckaert, Luk, Hendrik Opdebeeck, and László Zsolnai. *Frugality: Rebalancing Material and Spiritual Values in Economic Life*. New York: Peter Lang, 2008.

Brienza, Justin P., and Igor Grossmann. "Social Class and Wise Reasoning about Interpersonal Conflicts across Regions, Persons and Situations." *Proceedings of the Royal Society B: Biological Sciences* 284, no. 1869 (2017): 20171870. https://doi.org/10.1098/rspb.2017.1870.

Burt, Kate G., Hersh B. Luesse, Jennifer Rakoff, Andrea Ventura, and Marissa Burgermaster. "School Gardens in the United States: Current Barriers to Integration and Sustainability." *American Journal of Public Health* 108, no. 11 (2018): 1543–49. https://doi.org/10.2105/AJPH.2018.304674.

Cabin, Robert Jonathan. *Intelligent Tinkering: Bridging the Gap between Science and Practice*. Washington, DC: Island Press, 2011.

Campbell, Scott D. "Sustainable Development and Social Justice: Conflicting Urgencies and the Search for Common Ground in Urban and Regional Planning." *Michigan Journal of Sustainability* 1 (2013). https://doi.org/10.3998/mjs.12333712.0001.007.

Carlson, Allen. *Aesthetics and the Environment: The Appreciation of Nature, Art and Architecture*. New York: Routledge, 1999.

Carson, Rachel. *Silent Spring*. 40th anniversary ed. Boston: Mariner Books, 2002.

Carson, Rachel, and Dorothy Freeman. *Always, Rachel: The Letters of Rachel Carson and Dorothy Freeman, 1952–1964*. Edited by Martha E. Freeman. Boston: Beacon, 1995.

Centner, Terence J. *America's Blame Culture: Pointing Fingers and Shunning Restitution*. Durham, NC: Carolina Academic Press, 2008.

Cochrane, Willard. *Farm Prices: Myth and Reality*. Minneapolis: University of Minnesota Press, 1958.

Collins, Jim. *Good to Great: Why Some Companies Make the Leap and Others Don't*. New York: HarperBusiness, 2001.

Collins, Jim. "Level 5 Leadership: The Triumph of Humility and Fierce Resolve." *Harvard Business Review* 79, no. 1 (2001): 66–76.

Connor, Kathryn M., and Jonathan R. T. Davidson. "Development of a New Resilience Scale: The Connor-Davidson Resilience Scale (CD-RISC)." *Depression and Anxiety* 18, no. 2 (2003): 76–82.

Cortese, Anthony D. "The Critical Role of Higher Education in Creating a Sustainable Future." *Planning for Higher Education* 31, no. 3 (2003): 15–22.

Costanza, Robert, Lisa Graumlich, and W. Steffen. *Sustainability or Collapse? An Integrated History and Future of People on Earth*. Cambridge, MA: MIT Press, 2007.

Crocker, David A., and Toby Linden, eds. *Ethics of Consumption: The Good Life, Justice, and Global Stewardship*. Lanham, MD: Rowman & Littlefield, 1997.

Damasio, Antonio. *Descartes' Error: Emotion, Reason, and the Human Brain*. Repr. ed., London: Penguin Books, 2005.

Devall, Bill, and George Sessions. *Deep Ecology: Living as If Nature Mattered*. Salt Lake City: Gibbs Smith, 2007.

Diamond, Jared. *Collapse: How Societies Choose to Fail or Succeed*. Rev. ed. New York: Penguin Books, 2011.

Diener, Ed. "Subjective Well-Being: The Science of Happiness and a Proposal for a National Index." *American Psychologist* 55, no. 1 (2000): 34–43.

Dillard, Annie. *Pilgrim at Tinker Creek*. New York: Harper Perennial Modern Classics, 2013.

Dittmar, Helga, Rod Bond, Megan Hurst, and Tim Kasser. "The Relationship between Materialism and Personal Well-Being: A Meta-analysis." *Journal of Personality and Social Psychology* 107, no. 5 (2014): 879–924.

Dobson, Andrew. "Environment Sustainabilities: An Analysis and a Typology." *Environmental Politics* 5, no. 3 (1996): 401–28.

Doris, John M. "Persons, Situations, and Virtue Ethics." *Noûs* 32, no. 4 (1998): 504–30.

Duckworth, Angela. *Grit: The Power of Passion and Perseverance*. New York: Scribner, 2018.

Durlak, Joseph A., Roger P. Weissberg, Allison B. Dymnicki, Rebecca D. Taylor, and Kriston B. Schellinger. "The Impact of Enhancing Students' Social and Emotional Learning: A Meta-analysis of School-Based Universal Interventions." *Child Development* 82, no. 1 (2011): 405–32.

Ehrenfeld, John R. "The Roots of Sustainability." *MIT Sloan Management Review*, January 15, 2005.

Emmons, Robert A., and Michael E. McCullough, eds. *The Psychology of Gratitude*. Series in Affective Science. New York: Oxford University Press, 2004.

Epicurus. "Letter to Menoeceus." In *The Stoic and Epicurean Philosophers: The Complete Extant Writings of Epicurus, Epictetus, Lucretius [and] Marcus Aurelius*, edited by Whitney Jennings Oates. New York: Random House, 1940.

Farley, Heather M., and Zachary A. Smith. *Sustainability: If It's Everything, Is It Nothing?* 2nd ed. New York: Routledge, 2020.

Farnsworth, Ward. *The Practicing Stoic: A Philosophical User's Manual*. Jaffrey, NH: David R. Godine, 2018.

Fast, Nathanael J., and Larissa Z. Tiedens. "Blame Contagion: The Automatic Transmission of Self-Serving Attributions." *Journal of Experimental Social Psychology* 46, no. 1 (2010): 97–106.

Fath, Brian, Carly Dean, and Harald Katzmair. "Navigating the Adaptive Cycle: An Approach to Managing the Resilience of Social Systems." *Ecology and Society* 20, no. 2 (2015). https://doi.org/10.5751/ES-07467-200224.

Fesmire, Steven. *John Dewey and Moral Imagination: Pragmatism and Ethics*. Bloomington: Indiana University Press, 2003.

Franklin, Benjamin. *Autobiography and Other Writings*. Edited by Ormand Seavey. Oxford: Oxford University Press, 2009.

Gray, Barbara. "Conditions Facilitating Interorganizational Collaboration." *Human Relations* 38, no. 10 (1985): 911–36.

Greenleaf, Robert K. *Servant Leadership: A Journey into the Nature of Legitimate Power and Greatness*. New York: Paulist, 1977.

Grossmann, Igor, and Justin Brienza. "The Strengths of Wisdom Provide Unique Contributions to Improved Leadership, Sustainability, Inequality, Gross National Happiness, and Civic Discourse in the Face of Contemporary World Problems." *Journal of Intelligence* 6, no. 2 (2018): 22. https://doi.org/10.3390/jintelligence6020022.

Gunderson, Lance H., and C. S. Holling, eds. *Panarchy: Understanding Transformations in Human and Natural Systems*. Washington, DC: Island Press, 2001.

Hart, David Bentley. *The Hidden and the Manifest: Essays in Theology and Metaphysics*. Grand Rapids, MI: Eerdmans, 2017.

Hawken, Paul. *Blessed Unrest: How the Largest Social Movement in History Is Restoring Grace, Justice, and Beauty to the World*. Repr. ed. New York: Penguin Books, 2008.

Hayhoe, Katharine. *Saving Us: A Climate Scientist's Case for Hope and Healing in a Divided World*. New York: Atria / One Signal, 2021.

Heckman, James J., and Tim Kautz. "Fostering and Measuring Skills: Interventions That Improve Character and Cognition." NBER Working Papers. National Bureau of Economic Research, November 2013. https://ideas.repec.org/p/nbr/nberwo/19656.html.

Heckman, James J., and Tim Kautz. "Hard Evidence on Soft Skills." *Labour Economics* 19, no. 4 (2012): 451–64. https://doi.org/10.1016/j.labeco.2012.05.014.

Heinberg, Richard. *Welcome to the Great Unraveling: Navigating the Polycrisis of Environmental and Social Breakdown*. Corvallis, OR: Post Carbon Institute, 2023.

Hoffman, Martin. "Empathy, Justice, and the Law." In *Empathy: Philosophical and Psychological Perspectives*, edited by Amy Coplan and Peter Goldie, 230–54. New York: Oxford University Press, 2011.

Holland, John. "Studying Complex Adaptive Systems." *Journal of Systems Science and Complexity* 19 (2006): 1–8. https://doi.org/10.1007/s11424-006-0001-z.

Hopkins, Rob. *The Transition Handbook: From Oil Dependency to Local Resilience*. Cambridge: Green Books, 2014.

Hoyos-Valdés, Diana. "The Notion of Character Friendship and the Cultivation of Virtue." *Journal for the Theory of Social Behaviour* 48, no. 1 (2018): 66–82. https://doi.org/10.1111/jtsb.12154.

Hyde, Janet S., Elizabeth Fennema, and Susan J. Lamon. "Gender Differences in Mathematics Performance: A Meta-analysis." *Psychological Bulletin* 107, no. 2 (1990): 139–55.

Isaacson, Walter. *Benjamin Franklin: An American Life*. New York: Simon & Schuster, 2004.

Jacobs, Michael. "Sustainable Development as a Contested Concept." In *Fairness and Futurity: Essays on Environmental Sustainability and Social Justice*, edited by Andrew Dobson, 21–45. Oxford: Oxford University Press, 1999.

Jayawickreme, Eranda, Peter Meindl, Erik G. Helzer, R. Michael Furr, and William Fleeson. "Virtuous States and Virtuous Traits: How the Empirical Evidence regarding the Existence of Broad Traits Saves Virtue Ethics from the Situationist Critique." *Theory and Research in Education* 12, no. 3 (2014): 283–308. https://doi.org/10.1177/1477878514545206.

Kaplan, Rachel, and Ruby K. Blume. *Urban Homesteading: Heirloom Skills for Sustainable Living*. New York: Skyhorse, 2011.

Karelas, Andreas. *Climate Courage: How Tackling Climate Change Can Build Community, Transform the Economy, and Bridge the Political Divide in America*. Boston: Beacon, 2020.

Kashdan, Todd B., Melissa C. Stiksma, David J. Disabato, Patrick E. McKnight, John Bekier, Joel Kaji, and Rachel Lazarus. "The Five-Dimensional Curiosity Scale: Capturing the Bandwidth of Curiosity and Identifying Four Unique Subgroups of Curious People." *Journal of Research in Personality* 73 (2018): 130–49. https://doi.org/10.1016/j.jrp.2017.11.011.

Kastens, Kim A., Cathryn A. Manduca, Cinzia Cervato, Robert Frodeman, Charles Goodwin, Lynn S. Liben, David W. Mogk, Timothy C. Spangler, Neil A. Stillings, and Sarah Titus. "How Geoscientists Think and Learn." *Eos, Transactions, American Geophysical Union* 90, no. 31 (2009): 265–66.

Kautz, Tim, James J. Heckman, Ron Diris, Bas ter Weel, and Lex Borghans. "Fostering and Measuring Skills: Improving Cognitive and Non-cognitive Skills to

Promote Lifetime Success." OECD Education Working Papers 110, 2014. https://doi.org/10.3386/w20749.

Kimmerer, Robin Wall. *Braiding Sweetgrass: Indigenous Wisdom, Scientific Knowledge and the Teachings of Plants*. Minneapolis: Milkweed, 2020.

Kirchin, Simon, ed. *Thick Concepts*. Mind Association Occasional Series. Oxford: Oxford University Press, 2013.

Kramer, Roderick M., and Roy J. Lewicki. "Repairing and Enhancing Trust: Approaches to Reducing Organizational Trust Deficits." *Academy of Management Annals* 4, no. 1 (2010): 245–77.

Kraus, Michael W., Stéphane Côté, and Dacher Keltner. "Social Class, Contextualism, and Empathic Accuracy." *Psychological Science* 21, no. 11 (2010): 1716–23.

Kunstler, James Howard. *The Long Emergency: Surviving the End of Oil, Climate Change, and Other Converging Catastrophes of the Twenty-First Century*. New York: Grove, 2006.

Lakatos, Imry. "Falsification and the Methodology of Scientific Research Programmes." In *Criticism and the Growth of Knowledge*, edited by Imre Lakatos and Alan Musgrave, 91–196. Cambridge: Cambridge University Press, 1970.

Lao Tzu. *Tao Te Ching*. Translated by R. B. Blakney. New York: Signet, 2007.

Lear, Jonathan. *Radical Hope—Ethics in the Face of Cultural Devastation*. Cambridge, MA: Harvard University Press, 2008.

Leeuw, Sander van der, and Carl Folke. "The Social Dynamics of Basins of Attraction." *Ecology and Society* 26, no. 1 (2021): 33.

Lenton, Timothy M., Hermann Held, Elmar Kriegler, Jim W. Hall, Wolfgang Lucht, Stefan Rahmstorf, and Hans Joachim Schellnhuber. "Tipping Elements in the Earth's Climate System." *Proceedings of the National Academy of Sciences of the United States of America* 105 (2008): 1786–93. https://doi.org/10.1073/pnas.0705414105.

Leopold, Aldo. *A Sand County Almanac*. 1st Ballantine ed. New York: Sierra Club / Ballantine, 1970.

Lerch, Daniel. *Six Foundations for Building Community Resilience*. Corvallis, OR: Post Carbon Institute, 2015.

Lewicki, Roy J., and Chad Brinsfield. "Trust Repair." *Annual Review of Organizational Psychology and Organizational Behavior* 4, no. 1 (2017): 287–313.

Lynch, Michael P. *Know-It-All Society: Truth and Arrogance in Political Culture*. New York: Liveright, 2019.

Markovits, Daniel. *The Meritocracy Trap: How America's Foundational Myth Feeds Inequality, Dismantles the Middle Class, and Devours the Elite*. New York: Penguin, 2019.

Martin, Adrienne M. *How We Hope: A Moral Psychology*. Princeton, NJ: Princeton University Press, 2013.

McAfee, Andrew. *More from Less: The Surprising Story of How We Learned to Prosper Using Fewer Resources—and What Happens Next*. New York: Scribner, 2019.

McGeer, Victoria. "The Art of Good Hope." *Annals of the American Academy of Political and Social Science* 592, no. 1 (2004): 100–127.

McGerr, Michael. *A Fierce Discontent: The Rise and Fall of the Progressive Movement in America, 1870–1920*. Oxford: Oxford University Press, 2005.

McGinnis, Michael Vincent. *Science and Sensibility: Negotiating an Ecology of Place*. Oakland: University of California Press, 2016.

McKibben, Bill. *Falter: Has the Human Game Begun to Play Itself Out?* New York: Henry Holt, 2019.

McKibben, Bill. *Maybe One: A Case for Smaller Families*. New York: Plume, 1999.

Meadows, Donella H. *Thinking in Systems: A Primer*. Edited by Diana Wright. White River Junction, VT: Chelsea Green, 2008.

Meadows, Donella H., Dennis L. Meadows, Jørgen Randers, and William W. Behrens III. *The Limits to Growth: A Report for the Club of Rome's Project on the Predicament of Mankind*. New York: Universe Books, 1972.

Mills, Stephanie. "Population and the Planet: An Exchange." *New York Review of Books*, November 18, 2021.

Mischel, W., E. B. Ebbesen, and A. R. Zeiss. "Cognitive and Attentional Mechanisms in Delay of Gratification." *Journal of Personality and Social Psychology* 21, no. 2 (1972): 204–18.

Mischel, Walter, Yuichi Shoda, and Philip K. Peake. "The Nature of Adolescent Competencies Predicted by Preschool Delay of Gratification." *Journal of Personality and Social Psychology* 54, no. 4 (1988): 687–96.

Mocker, Donald W., and George E. Spear. *Lifelong Learning: Formal, Nonformal, Informal, and Self-Directed*. Information Series 241. Columbus, OH: National Center, 1982.

Naess, Arne. "The Shallow and the Deep, Long-Range Ecology Movement. A Summary." *Inquiry: An Interdisciplinary Journal of Philosophy* 16, no. 1–4 (1973): 95–100.

Nickerson, Raymond S. "Confirmation Bias: A Ubiquitous Phenomenon in Many Guises." *Review of General Psychology* 2, no. 2 (1998): 175–220.

Nietzsche, Friedrich. *The Gay Science: With a Prelude in Rhymes and an Appendix of Songs*. Translated by Walter Kaufmann. New York: Vintage Books, 1974.

Nightingale, Andrea, Siri Eriksen, Marcus Taylor, Timothy Forsyth, Mark Pelling, Andrew Newsham, Emily Boyd, et al. "Beyond Technical Fixes: Climate Solutions and the Great Derangement." *Climate and Development* 12 (2020): 343–52. https://doi.org/10.1080/17565529.2019.1624495.

Norton, Bryan G. *Sustainable Values, Sustainable Change: A Guide to Environmental Decision Making*. Chicago: University of Chicago Press, 2015.

Nussbaum, Martha C. *Creating Capabilities: The Human Development Approach*. Repr. ed. Cambridge, MA: Belknap Press of Harvard University Press, 2013.

Nussbaum, Martha C. *The Fragility of Goodness: Luck and Ethics in Greek Tragedy and Philosophy*. 2nd ed. Cambridge: Cambridge University Press, 2001.

Nussbaum, Martha C. *Upheavals of Thought: The Intelligence of Emotions*. Cambridge: Cambridge University Press, 2001.

Oliver, Mary. *Evidence: Poems*. Repr. ed. Boston: Beacon, 2010.

O'Neill, Brian C., Elmar Kriegler, Kristie L. Ebi, Eric Kemp-Benedict, Keywan Riahi, Dale S. Rothman, and Bas J. van Ruijven. "The Roads Ahead: Narratives for Shared Socioeconomic Pathways Describing World Futures in the 21st Century." *Global Environmental Change* 42 (2017): 169–80. https://doi.org/10.1016/j.gloenvcha.2015.01.004.

Oreskes, Naomi, and Erik Conway. *The Collapse of Western Civilization: A View from the Future*. New York: Columbia University Press, 2014.

Ortman, Jennifer, Victoria Velkoff, and Howard Hogan. "An Aging Nation: The Older Population in the United States." United States Census Bureau, 2014. https://www.census.gov/library/publications/2014/demo/p25-1140.html.

Pennock, Robert T. *An Instinct for Truth: Curiosity and the Moral Character of Science*. Cambridge, MA: MIT Press, 2019.

Peterson, Christopher, and Martin Seligman. *Character Strengths and Virtues: A Handbook and Classification*. New York: American Psychological Association and Oxford University Press, 2004.

Piasecki, Bruce. *Doing More with Less: The New Way to Wealth*. Hoboken, NJ: Wiley, 2012.

Prinz, Jesse. "Against Empathy." *Southern Journal of Philosophy* 49, no. s1 (2011): 214–33.

Putnam, Hilary. *The Many Faces of Realism*. La Salle, IL: Open Court, 1987.

Putnam, Robert D. *Bowling Alone: The Collapse and Revival of American Community*. New York: Simon & Schuster, 2000.

Putnam, Robert D., and Shaylyn Romney Garrett. *The Upswing: How America Came Together a Century Ago and How We Can Do It Again*. New York: Simon & Schuster, 2020.

Rice, Christopher M. "Defending the Objective List Theory of Well-Being." *Ratio* 26, no. 2 (2013): 196–211. https://doi.org/10.1111/rati.12007.

Rittel, Horst W. J., and Melvin M. Webber. "Dilemmas in a General Theory of Planning." *Policy Sciences* 4, no. 2 (1973): 155–69.

Roberts, Robert C., and W. Jay Wood. *Intellectual Virtues: An Essay in Regulative Epistemology*. Oxford: Oxford University Press, 2007.

Robinson, Kim Stanley. *The Ministry for the Future*. New York: Orbit, 2020.

Rockström, Johan, Will Steffen, Kevin Noone, Asa Persson, F. Stuart Chapin 3rd, Eric F. Lambin, Timothy M. Lenton, et al. "A Safe Operating Space for Humanity." *Nature* 461, no. 7263 (2009): 472–75.

Rosenberg, Joshua. "Teaching Empathy in Law School." *University of San Francisco Law Review* 36, no. 3 (2002), 625–57.

Ross, Alec. *The Industries of the Future*. New York: Simon & Schuster, 2017.

Ruiter, Robert, Loes Kessels, Gjalt-Jorn Peters, and Gerjo Kok. "Sixty Years of Fear Appeal Research: Current State of the Evidence." *International Journal of Psychology* 49 (2014): 63–70. https://doi.org/10.1002/ijop.12042.

Sabatier, Paul, Will Focht, Mark Lubell, Zev Trachtenberg, Arnold Vedlitz, and Marty Matlock, eds. *Swimming Upstream: Collaborative Approaches to Watershed Management*. Cambridge, MA: MIT Press, 2006.

Saphire-Bernstein, Shimon, and Shelley E. Taylor. *Close Relationships and Happiness*. Oxford: Oxford University Press, 2013.

Schein, Edgar H. "How Can Organizations Learn Faster? The Challenge of Entering the Green Room." *MIT Sloan Management Review*, January 15, 1993.

Schor, Juliet. *Plenitude: The New Economics of True Wealth*. New York: Penguin, 2010.

Seligman, Martin E. P. *Flourish: A Visionary New Understanding of Happiness and Well-Being*. New York: Atria Books, 2012.

Sen, Amartya. "Capability and Well-Being." In *The Quality of Life*, edited by Martha Nussbaum and Amartya Sen, 30–53. Oxford: Clarendon, 1993.

Senge, Peter M. *The Fifth Discipline: The Art and Practice of the Learning Organization*. Rev. and updated ed. New York: Doubleday, 2006.

Shellenberger, Michael. *Apocalypse Never: Why Environmental Alarmism Hurts Us All*. New York: Harper, 2020.

Shellenberger, Michael, and Ted Nordhouse. "The Death of Environmentalism: Global Warming Politics in a Post Environmental World." 2004. www.thebreakthrough.org/images/Death_of_Environmentalism.pdf.

Singer, Peter. *One World: The Ethics of Globalization*. New Haven, CT: Yale University Press, 2002.

Slater, Michael D. "Reinforcing Spirals: The Mutual Influence of Media Selectivity and Media Effects and Their Impact on Individual Behavior and Social Identity." *Communication Theory* 17, no. 3 (2007): 281–303.

Snyder, C. R. "Conceptualizing, Measuring, and Nurturing Hope." *Journal of Counseling & Development* 73, no. 3 (1995): 355–60.

Solomon, Robert C. *The Passions*. Garden City, NY: Anchor / Doubleday, 1976.

Solomon, Robert C., and Fernando F. Flores. *Building Trust: In Business, Politics, Relationships, and Life*. New York: Oxford University Press, 2001.

Sotomayor, Sonia. "A Latina Judge's Voice." *Berkeley La Raza Law Journal* 13, no. 1 (2002): 87–93.

Sotomayor, Sonia. *My Beloved World*. Repr. ed. New York: Vintage Books, 2014.

Stamos, Angelos, Efthymios Altsitsiadis, and Siegfried Dewitte. "Investigating the Effect of Childhood Socioeconomic Background on Interpersonal Trust: Lower Childhood Socioeconomic Status Predicts Lower Levels of Trust." *Personality and Individual Differences* 145 (2019): 19–25.

Steffen, Will, Katherine Richardson, Johan Rockström, Sarah E. Cornell, Ingo Fetzer, Elena M. Bennett, Reinette Biggs, et al. "Planetary Boundaries: Guiding Human Development on a Changing Planet." *Science* 347, no. 6223 (2015).

Steffen, Will, Johan Rockström, Katherine Richardson, Timothy M. Lenton, Carl Folke, Diana Liverman, Colin P. Summerhayes, et al. "Trajectories of the Earth System in the Anthropocene." *Proceedings of the National Academy of Sciences* 115, no. 33 (2018): 8252–59.

Sterling, Stephen. "Learning for Resilience, or the Resilient Learner? Towards a Necessary Reconciliation in a Paradigm of Sustainable Education." *Environmental Education Research* 16, no. 5–6 (2010): 511–28.

Stonehouse, Paul, and William Throop. "Coping with Climate Despair: Cultivating the Skills of Hope and Tranquil Resolve." *Journal of Sustainability Education* 28 (2023).

Stroh, David Peter. *Systems Thinking for Social Change: A Practical Guide to Solving Complex Problems, Avoiding Unintended Consequences, and Achieving Lasting Results*. White River Junction, VT: Chelsea Green, 2015.

Sunstein, Cass R. "The Law of Group Polarization." *Journal of Political Philosophy* 10, no. 2 (2002): 175–95.

Tannen, Deborah. *The Argument Culture: Stopping America's War of Words*. New York: Ballantine Books, 1999.

Thomashow, Mitchell. *The Nine Elements of a Sustainable Campus*. Cambridge, MA: MIT Press, 2016.

Thompson, Allen. "Radical Hope for Living Well in a Warmer World." *Journal of Agricultural and Environmental Ethics* 23, no. 1–2 (2009): 43–55.

Thompson, Ashley E., and Daniel Voyer. "Sex Differences in the Ability to Recognise Non-verbal Displays of Emotion: A Meta-analysis." *Cognition and Emotion* 28, no. 7 (2014): 1164–95. https://doi.org/10.1080/02699931.2013.875889.

Throop, William. "Flourishing in the Age of Climate Change: Finding the Heart of Sustainability." *Midwest Studies in Philosophy* 40 (2016): 296–314. https://doi.org/10.1111/misp.12062.

Throop, William. "From Environmental Advocates to Sustainability Entrepreneurs: Rethinking a Sustainability Focused General Education Program." In *Sustainability in Higher Education: Stories and Strategies for Transformation*, edited by Peggy F. Barlett and Geoffrey W. Chase, 81–88. Cambridge, MA: MIT Press, 2013.

Throop, William. "Frugality and Resilience: A Pragmatist Meditation." In *Pragmatist and American Philosophical Perspectives on Resilience*, edited by Kelley A. Parker and Heather E. Keith, 247–60. New York: Rowman & Littlefield, 2019.

Throop, William. "Learning Our Way toward Resilience." In *The Community Resilience Reader: Essential Resources for an Era of Upheaval*, edited by Daniel Lerch, 247–60. Washington, DC: Island Press, 2017. https://doi.org/10.5822/978-1-61091-861-9_15.

Throop, William, and Matt Mayberry. "Leadership for the Sustainability Transition." *Business and Society Review* 122, no. 2 (2017): 221–50. https://doi.org/10.1111/basr.12116.

Thunberg, Greta. *No One Is Too Small to Make a Difference*. London: Penguin, 2019.

Tiberius, Valerie. "Value Commitments and the Balanced Life." *Utilitas* 17, no. 1 (2005): 24–45.

Tiberius, Valerie, and Alexandra Plakias. "Well-Being." In *The Moral Psychology Handbook*, edited by John Michael Doris, 402–32. New York: Oxford University Press, 2010.

Twenge, Jean M., A. Bell Cooper, Thomas E. Joiner, Mary E. Duffy, and Sarah G. Binau. "Age, Period, and Cohort Trends in Mood Disorder Indicators and Suicide-Related Outcomes in a Nationally Representative Dataset, 2005–2017." *Journal of Abnormal Psychology* 128, no. 3 (2019): 185–99. https://doi.org/10.1037/abn0000410.

Walker, Brian, and Jacqueline Meyers. "Thresholds in Ecological and Social-Ecological Systems: A Developing Database." *Ecology and Society* 9, no. 2 (2004): 3. http://www.ecologyandsociety.org/vol9/iss2/art3/.

Walker, Brian, and David Salt. *Resilience Practice: Building Capacity to Absorb Disturbance and Maintain Function*. Washington, DC: Island Press, 2012.

Walker, Kenny, and Lynda Walsh. "'No One Yet Knows What the Ultimate Consequences May Be': How Rachel Carson Transformed Scientific Uncertainty into a Site for Public Participation in *Silent Spring*." *Journal of Business and Technical Communication* 26, no. 1 (2012): 3–34. https://doi.org/10.1177/1050651911421122.

Wallace-Wells, David. *The Uninhabitable Earth: Life after Warming*. New York: Tim Duggan Books, 2019.

Wason, P. C. "On the Failure to Eliminate Hypotheses in a Conceptual Task." *Quarterly Journal of Experimental Psychology* 12, no. 3 (1960): 129–40.

Webster, Noah J., Kristine J. Ajrouch, and Toni C. Antonucci. "Sociodemographic Differences in Humility: The Role of Social Relations." *Research in Human Development* 15, no. 1 (2017): 1–22. https://doi.org/10.1080/15427609.2017.1414670.

West, Darrell M. *The Future of Work: Robots, AI, and Automation*. Washington, DC: Brookings Institution, 2018.

Westacott, Emrys. *The Wisdom of Frugality: Why Less Is More—More or Less*. Princeton, NJ: Princeton University Press, 2016.

Weziak-Bialowolska, Doroto, Piotr Bialowolski, Matthew T. Lee, Ying Chen, Tyler J. VanderWeele, and Eileen McNeely. "Psychometric Properties of Flourishing Scales from a Comprehensive Well-Being Assessment." *Frontiers in Psychology* 12 (2021). https://doi.org/10.3389/fpsyg.2021.652209.

Whitcomb, Dennis, Heather Battaly, Jason Baehr, and Daniel Howard-Snyder. "Intellectual Humility: Owning Our Limitations." *Philosophy and Phenomenological Research* 94, no. 3 (2017): 509–39. https://doi.org/10.1111/phpr.12228.

Williams, Terry Tempest. *Finding Beauty in a Broken World*. New York: Vintage Books, 2009.

Williston, Byron. *The Anthropocene Project: Virtue in the Age of Climate Change*. Oxford: Oxford University Press, 2015.

Wilson, David Sloan, Steven C. Hayes, Anthony Biglan, and Dennis D. Embry. "Evolving the Future: Toward a Science of Intentional Change." *Behavioral and Brain Sciences* 37, no. 4 (2014): 395–416. https://doi.org/10.1017/S0140525X13001593.

Witkowski, Terrence H. "A Brief History of Frugality Discourses in the United States." *Consumption Markets & Culture* 13, no. 3 (2010): 235–58. https://doi.org/10.1080/10253861003786975.

Wondolleck, Julia Marie, and Steven Lewis Yaffee. *Making Collaboration Work: Lessons from Innovation in Natural Resource Management*. Washington, DC: Island Press, 2000.

Woolley, John, Michael McGinnis, and Julie Kellner. "The California Watershed Movement: Science and the Politics of Place." *Natural Resources Journal* 42, no. 1 (2002): 133–83.

Worthington, Everett, Don E. Davis, and Joshua N. Hook. *Handbook of Humility: Theory, Research, and Applications*. New York: Routledge, 2016.

Yates, Joshua, and James Davison Hunter, eds. *Thrift and Thriving in America: Capitalism and Moral Order from the Puritans to the Present*. New York: Oxford University Press, 2011.

Yeats, William Butler. "The Second Coming." 1919. Text/html. Poetry Foundation. https://www.poetryfoundation.org/poems/43290/the-second-coming.

Index

Figures and boxes are indicated by "*f*" and "**b**" following page numbers.